CLIMATE CHANGE AND ITS CAUSES, EFFECTS AND PREDICTION

WATER UTILITY APPROACHES AND RESPONSES TO CLIMATE CHANGE VULNERABILITY

Climate Change and its Causes, Effects and Prediction

Additional books in this series can be found on Nova's website under the Series tab.

Additional E-books in this series can be found on Nova's website under the E-book tab.

Water Resource Planning, Development and Management

Additional books in this series can be found on Nova's website under the Series tab.

Additional E-books in this series can be found on Nova's website under the E-book tab.

CLIMATE CHANGE AND ITS CAUSES, EFFECTS AND PREDICTION

WATER UTILITY APPROACHES AND RESPONSES TO CLIMATE CHANGE VULNERABILITY

James A. Williams
and
Caroline E. Jones
EDITORS

Nova Science Publishers, Inc.
New York

For permission to use material from this book please contact us:
Telephone 631-231-7269; Fax 631-231-8175
Web Site: http://www.novapublishers.com

Library of Congress Cataloging-in-Publication Data

ISBN 978-1-61942-784-6

Published by Nova Science Publishers, Inc. † New York

CONTENTS

PREFACE

Climate change poses a variety of challenges for water management, and there is a need to develop methods for understanding and managing risk. While much has been written about the projected impacts of climate change at the continental or regional scale, scientists are quick to caution decision makers about using projections based on global circulation models (GCMs) for local decision making. This uncertainty about specific impacts on local systems has raised concern about the ability of water resource managers to plan for climate and hydrological changes at the local scale, and has spurred recent activity to develop methods for understanding vulnerabilities, including how to downscale climate models. This book examines and documents the steps taken by some of the leading utilities in an attempt to identify the emergent characteristics of water utility climate change vulnerability assessments.

Chapter 1- Climate change poses a variety of challenges for water management, and there is a need to develop methods for understanding and managing risk. While much has been written about the projected impacts of climate change at the continental or regional scale, scientists are quick to caution decision makers about using projections based on global circulation models (GCMs) for local decision making. This 'uncertainty' about specific impacts on local systems has raised concern about the ability of water resource managers to plan for climatic and hydrological changes at the local scale, and has spurred recent activity to develop methods for understanding vulnerabilities, including how to downscale climate models.

This study examines and documents the steps taken by some of the leading utilities in an attempt to identify the emergent characteristics of water utility climate change vulnerability assessments. By examining the approaches taken and articulating the steps, information, and judgments needed for such decision making, we hope to contribute to the collaborative problem solving among the user and research communities who are working to further refine and validate such procedures.

Chapter 2- Concern about the potential effects of climate change on the quantity, quality, timing, and demand for water is growing. The implications of climate change for long-lived, capital-intensive water infrastructure is a particular concern. In 2009, the U.S. Environmental Protection Agency sponsored the *First National Expert and Stakeholder Workshop on Water Infrastructure Sustainability and Adaptation to Climate Change* (U.S. EPA, 2009a). This workshop highlighted a need for improved information and tools to help water utilities better understand and manage the risk of future impacts due to climate change. Many utilities are already addressing this challenge. One finding of this workshop was that compiling case studies of current water utility activities related to climate change adaptation would be useful to help those in the water sector learn from each other.

This report presents case studies describing the approaches taken by four water utilities in the United States to assess their vulnerability to climate change. The report is not intended to be a comprehensive listing of assessment approaches or utilities conducting vulnerability assessments. Nor does this report represent the full range of potential approaches for assessing and managing climate risk. Rather, the purpose of this report is to illustrate a range of approaches that selected water utilities have taken to understand and respond to climate risk.

In: Water Utility Approaches and Responses... ISBN: 978-1- 61942-784-6
Editors:J. A. Williams and C. E. Jones:

Chapter 1

CLIMATE CHANGE VULNERABILITY ASSESSMENTS: A REVIEW OF WATER UTILITY PRACTICES*

Office of Water Environmental Protection Agency

ACKNOWLEDGMENTS

This report was prepared for EPA by Stratus Consulting Inc., Boulder, CO, under subcontract to Abt Associates, Inc., under EPA Contract EP-C-07-023, Work Assignment 2-13. Karen Metchis served as Technical Project Officer and principal reviewer of this document.

EPA wishes to acknowledge the principle authors, Jason M. Vogel and Joel B. Smith for their diligence and expertise in preparing this report. Further thanks go to EPA ORD project officer, Jill Neal, and ORD scientists Jeffrey Yang, Ph.D., and Thomas Johnson, Ph.D., for their comments and suggestions in helping shape this report.

This report represents one product resulting from the "First National Expert Workshop on Water Utilities and Climate Change," held January 20-21, 2009, in Washington, D.C. Acknowledgement also goes to Elizabeth Corr,

* This is an edited, reformatted and augmented version of the United States Environmental Protection Agency, 800-R-10-001, dated August 2010.

Robert Cantilli, EPA Office of Water, and Jeff Yang, Ph.D., ORD, for their collaboration in formulating the workshop.

1. Executive Summary

Climate change poses a variety of challenges for water management, and there is a need to develop methods for understanding and managing risk. While much has been written about the projected impacts of climate change at the continental or regional scale, scientists are quick to caution decision makers about using projections based on global circulation models (GCMs) for local decision making. This 'uncertainty' about specific impacts on local systems has raised concern about the ability of water resource managers to plan for climatic and hydrological changes at the local scale, and has spurred recent activity to develop methods for understanding vulnerabilities, including how to downscale climate models.

This study examines and documents the steps taken by some of the leading utilities in an attempt to identify the emergent characteristics of water utility climate change vulnerability assessments. By examining the approaches taken and articulating the steps, information, and judgments needed for such decision making, we hope to contribute to the collaborative problem solving among the user and research communities who are working to further refine and validate such procedures.

The study describes the activities of eight water utilities who have conducted climate vulnerability assessments: East Bay Municipal Utility District, City of Boulder Utilities Division, Denver Water, Massachusetts Water Resources Authority, New York City Department of Environmental Protection, Portland Water Bureau, Lower Colorado River Authority, and Seattle Public Utilities.

The following general observations can be made. First, while the intention was to evaluate both drinking water and wastewater utilities in the systems evaluated, most vulnerability assessment work focused on water quantity and water demand and in only one case on water quality.

Second, while utility managers typically possess expertise about their systems, their hydrological and management models, and the local hydrology, they have more limited access to climate change information. Utilities are able to examine impacts on their operations due to changes in climate-sensitive variables (e.g., flow, demand), however, in many cases, in order to understand the changes in climate that affect those variables, utilities are engaging outside climate expertise.

Third, in each case study presented in this report, utilities are compensating for the 'uncertainty' about climate change by evaluating a range of scenarios or models, and in most cases the resulting information is used to test the robustness of existing decision making or planning against different future, plausible scenarios.

Climate change is a complex issue, and more work is needed to establish reliable practices for incorporating climate change into water utility decisions and planning. At the same time, utilities appear to have benefited from their various efforts to understand their potential vulnerabilities and to evaluate long term planning options. Despite the uncertainty, there are reasonable and prudent steps utilities and other water managers can take to better understand and manage climate risk.

2. Introduction

For many decades, water utilities have proactively assessed the ability of their systems to provide a reliable supply of drinking water or adequate wastewater services under assumptions of population change, municipal expansion (e.g., annexation), technological innovation, and changes in regulations. These were the foreseeable challenges utilities addressed to ensure that ratepayers received high quality water services at minimal cost.

Historically, utilities assumed stationarity of climate in their water resource planning – the idea that natural systems fluctuate within an unchanging envelope of variability. Recently, however, this assumption has been challenged, forcing managers to rethink future water resources planning and management with regard to climate change (Milly et al., 2008). In response, over the last decade, water utilities have expanded their risk assessment efforts to address changing climate conditions.

Generally speaking, vulnerability assessments done by water utilities are grounded in a thorough understanding of their water system. On a practical level, this means that vulnerability studies use a variety of tools and models that have been developed and refined to reflect the local hydrology, climate, infrastructure, operations, and demands that are unique to a particular water utility. Consequently, no single tool currently exists that can comprehensively address the vulnerability of diverse water systems to climate change. For example, most utilities have operational or management models that estimate the day-to-day operation of their water or wastewater systems. By necessity, these operational models are either unique to a particular utility (e.g., a proprietary system model) or in some cases, customized versions of a tool used by many utilities (e.g., the Water Evaluation and Planning, or WEAP, system model).

The purpose of this report is to illustrate the approaches used by water utilities to assess their vulnerability to climate change. This is a review of best practices in this emerging effort across the industry for the purpose of informing utilities considering engaging in this issue about the various methods used by their peers. The report does not judge or evaluate the efforts of utilities or the merits of different vulnerability assessment methods, but describes the efforts of eight utilities as broad approaches, tools, and methods worthy of consideration by other utilities depending upon their needs, available resources, and other factors.

In this report we examined the ways in which several water utilities evaluated their potential vulnerability to climate change. Utilities included in this study were identified from published studies on water resources and climate change, including six Water Research Foundation reports (Strange et al., 2009; WRF, 201 0a, 201 0b, 201 0c, 201 0d, 201 0f), two reports by the Water Utility Climate Alliance (Barsugli, et al., 2009; Means et al., 2010), and participants in the ongoing Joint Front Range Climate Change Vulnerability Study (WRF, 201 0e). We also included utilities selected from the list of participants in two recent workshops focused on water resources and climate change: the National Drinking Water Advisory Council"s Climate Ready Water Utilities Working Group and U.S. Environmental Protection Agency"s (EPA's) First National Expert and Stakeholder Workshop on Water Infrastructure Sustainability and Adaptation to Climate Change (U.S. EPA, 2009).[1] In total, we identified 50 water utilities as potential candidates for in-depth case studies. Initial review, however, revealed that many of the selected utilities were interested in tracking the climate change issue and learning what others had done, but had not yet engaged in their own climate change vulnerability assessments. If this was the case, the utility was eliminated from the initial list for further study.

We identified five utilities that had published reports on their own climate change vulnerability work, and another five utilities that appeared to have conducted vulnerability work but had not published the modeling results. Because of time and resource limitations, the study was limited to eight of these 10 utilities. The two utilities not included were Inland Empire Utilities

Agency and King County Wastewater Treatment Division. The eight utilities included in the assessment included:

- East Bay Municipal Utility District (EBMUD)
  - A water supply and wastewater utility serving 1.3 million customers.
- City of Boulder Utilities Division
  - A water supply and wastewater utility serving 113,000 customers.
- Denver Water
  - A water supply utility serving 1.3 million customers.
- Massachusetts Water Resources Authority (MWRA)
  - A water supply and wastewater utility serving 2.2 million customers.
- New York City Department of Environmental Protection (NYCDEP)
  - A water supply and wastewater utility serving 9.2 million customers.
- Portland Water Bureau (PWB)
  - A water supply utility serving 860,000 customers.
- Lower Colorado River Authority (LCRA)
  - A conservation and reclamation district that manages water supply along a 600- mile stretch of the Colorado River in Texas, operates six dams, and helps communities plan and coordinate their water and wastewater needs.
- Seattle Public Utilities (SPU)
  - A water supply utility serving 1.35 million customers.

Finally, we studied these eight utilities' climate change vulnerability analyses in depth to identify tools and approaches used by those water utilities in their assessments. The remainder of this report synthesizes the insights from the analysis of these eight utilities in the following sections: approaches to assessing climate change vulnerability, sources of climate information, modeling changes in water resources, summary, and recommendations for further study.

3. Approaches to Assessing Climate Change Vulnerability

Most water managers engaged in climate vulnerability analyses have a strong technical understanding of their water systems, including local hydrology, historical operating conditions, and standard operational practices.[2] However, they typically are not climatologists and have limited experience assessing risks from climate change. As a result, a water utility can either engage outside expertise to do a sophisticated exploration of the implications of climate change or they can make do with the information and capabilities readily available. Utilities that choose not to engage outside experts are generally limited to analyses based on their internal understanding of their water system and/or sensitivity analyses to assess what hypothetical climate changes would mean for their water system. In contrast, utilities that engage outside expertise can generate climate projections specific to their watershed. This more resource- intensive approach enables computationally sophisticated vulnerability analyses.

Approaches to assessing climate change risks are generally classified as either "top-down" modeling assessments or "bottom-up" threshold analyses (e.g., Miller and Yates, 2006; Freas et al., 2008; Stratus Consulting and MWH Global, 2009). This categorization is useful, and, generally speaking, utilities choose one of these two approaches to initiate their climate vulnerability assessment. However, it should be noted that these two approaches are not mutually exclusive and utilities often incorporate elements of both, either in series or in parallel. The choice of one approach over another, however, has significant implications for study design and resource needs, making them useful as categorical descriptions. The following utilities initially used a top-down assessment:

- City of Boulder Utilities Division
- Denver Water
- MWRA
- NYCDEP
- PWB
- SPU

The following utilities initially used a bottom-up assessment:

- EBMUD
- LCRA

In general, top-down assessments are model and data driven. They often are more time and resource intensive than bottom-up assessments, sometimes requiring expertise beyond the capacity of many utilities. In general, bottom-up assessments are driven by knowledge of the utility system itself. Such assessments often focus on defining critical system thresholds or specific decisions that may be sensitive to climate change. This can often be done qualitatively, and the necessary expertise often resides within the utility. As suggested above, these two approaches are not mutually exclusive.

3.1. Top-down Modeling Assessments

Six utilities in our study initiated their vulnerability assessments with a top-down approach: Boulder Utilities (Smith et al., 2009), Denver Water (Laurna Kaatz, personal communication, April 8, 2009), MWRA (Stephen Estes-Smargiassi, personal communication, February 26, 2010), NYCDEP (2008; Major et al., 2007), PWB (Palmer and Hahn, 2002), and SPU (Palmer, 2007). This method projects future climate conditions specific to the watershed of concern and is the most computationally and resource intensive of the analysis approaches used by utilities. Because of this, the top-down modeling assessments done by the utilities examined for this report were often completed by the water utility in conjunction with an academic institution or consulting firm with expertise in climate modeling. For example, SPU and PWB partnered with the University of Washington's Climate Impacts Group (UW-CIG), NYCDEP partnered with the City University of New York Institute for Sustainable Cities (CISC) and Columbia University's Center for Climate Systems Research (CCSR), LCRA used CH2M Hill as a consultant, and the City of Boulder Utilities Division engaged Stratus Consulting and Hydrosphere (now AMEC).

Top-down assessments can involve relying on historic hydrology, using paleoclimate records, engaging in literature reviews, or they can be done using climate projections. All these options are discussed in more detail in Section 3 (Sources of Climate Information). By far, the most common form of top-down assessment involves generating downscaled climate projections for a utility"s local watershed using global climate model (GCM) output. To do this, the analysis generally includes assumptions about future emission levels (i.e., global socioeconomic development scenarios), the effects of those emission levels on global climate (e.g., temperature and precipitation), the translation of global climate effects to the region or watershed, and the specification of climate changes on hydrology (e.g., streamflow), as well as operational, management, and demand models.

All of the utilities we studied in depth assessed a number of climate variables; however, the most common climate variables of interest were those that fed into their watershed-specific hydrology models or system-specific management or operations models. Most often, this included temperature and precipitation data or streamflow data at the spatial and temporal scales of each individual utility's hydrology, management, operations, or other models.[3] Therefore, to make meaningful comparisons in this study, we focused our investigation of the different modeling approaches to develop projections of temperature and precipitation. It is worth noting, however, that some utilities also investigated issues of special concern for their water system. For example, NYCDEP was concerned about source water quality, especially due to heavy precipitation events and resulting turbidity, and MWRA was concerned about likely impacts on the safe yield of its sources.

Top-down modeling has three important subcategories: scenario analyses, sensitivity analyses, and paleocli mate data or historic climate observations to define temperature and precipitation patterns for water system planning purposes. Each of these subcategories is described in detail in the following sections.

3.1.1. Scenario analyses

The scenario approach to a top-down assessment begins by defining at least one, but more commonly two or three plausible greenhouse gas (GHG) emission futures that merit consideration by a water utility. A scenario analysis then works step by step through GCMs, downscaling those results,[4] and finally using those results in water system specific hydrology, demand, operational, and/or management models as appropriate to investigate the vulnerability of a water system to each scenario. Section 3.4 (Climate Projections) discusses these individual steps in greater detail. In many cases, scenarios are developed to capture a wide range of plausible futures by incorporating an extensive range of models, emissions scenarios, and projected demands. As an example, the SPU study applied the scenario analysis technique and used GCMs and socioeconomic projections to assess a "middle of the road," a "warmer and wetter," and a "warmer and drier" scenario. MWRA's approach used the outcome of all GCMs to estimate the probability distributions of likely changes in temperature and precipitation and then utilized those distributions in the statistical downscaling. This methodology, developed at NCAR, avoids using a scenario tied to a single GCM.

3.1.2. Sensitivity analyses

Another form of the top-down approach uses a sensitivity analysis, incorporating the use of incremental changes in climate such as 1°, 2°, and 3°C (1.8°, 3.6°, and 5.4°F) annual temperature increases combined with +/- 0%, 10%, or 20% annual changes in precipitation.[5] PWB used this type of sensitivity analysis to bracket a plausible range of climate-altered future hydrology. But utilities can also engage in a sensitivity analysis, as was done by EBMUD, with a bottom-up approach to avoid the computationally and resource intensive steps of a scenario analysis. A sensitivity analysis does not require using GHG emissions projections, using climate model projections, or downscaling climate model output.

3.1.3. Paleoclimate or historic analyses

A third top-down approach incorporates paleoclimate studies or historic climate observations to define temperature and precipitation patterns for water system planning purposes. The method often defines a worst case scenario (e.g., three consecutive years of the drought of record) or examines the water system effects of droughts in the paleorecord outside of observed variability. While the City of Boulder Utilities Division, Denver Water, and EBMU D used paleoclimate studies to complement their climate change vulnerability studies, none of them relied exclusively on paleocli mate or conservative extrapolations from historic data.[6]

3.2. Bottom-up Threshold Analyses

The bottom-up approach to climate change vulnerability assessments is grounded in knowledge of the water system itself. Using this approach, a qualitative system assessment is done to determine which system components are potentially vulnerable to change. The results of the assessment can be used to focus further study on specific impacts of concern.

EBMUD, which used this approach, stated, “In a ‘Bottom-Up’ approach, the most critical vulnerabilities of the District’s water supply system are identified, the causes of those vulnerabilities are articulated, and then steps are taken to better address and solve the vulnerability in the face of climatic uncertainty" (EBMUD, 2009, p. 4-16). They noted their concept of a bottom-up approach is adapted from the AwwaRF (now the Water Research Foundation) publication, *Climate Change and Water Resources: A Primer for Municipal Water Providers* (Miller and Yates, 2006).

More specifically, the EBMUD research approach involved a portfolio evaluation that first identified potential portfolio components (e.g., new reservoirs, expanded reservoir storage, increased conservation, conjunctive use, water reclamation, desalination, interbasin transfers), screened those components for technical, environmental, and economic feasibility, and then constructed alternate portfolios of multiple components that could meet projected demands (e.g., increased conservation and conjunctive use, or water reclamation and interbasin transfers). Then, a preliminary portfolio analysis was conducted using a combination of the WEAP system model and the district"s EBMUDsim model – known collectively as the "W-E model." Portfolios that performed poorly under current hydrological conditions were eliminated and the remaining portfolios were subjected to detailed analyses under anticipated climate change conditions using the W-E model. By screening out portfolios that perform poorly under current conditions, EBMUD implicitly limited options to those that could be implemented under current conditions without significant reduction of reliability, but which presumably would improve the system under climate change relative to the current system. By comparing the climate change scenarios to a baseline scenario, the sensitivity of the EBM UD water system to each component was assessed to identify critical vulnerabilities and identify portfolios that addressed those vulnerabilities.

LCRA also used a bottom-up approach, and in CH2M Hill's report for the authority, the threshold approach was defined in this way: "This threshold approach identifies system components that are dependent on the status of climate variables (e.g., precipitation, temperature, etc.) and the overall system risk to climate change, resulting in a preliminary risk assessment based on the professional judgment of experts who know the system and the planning area" (CH2M Hill, 2008, p. 3-2). This approach is a qualitative or semi-quantitative analysis that consisted of (1) identifying the climate variables of importance and exploring the sensitivity of LCRA to these variables; (2) determining water system responses to a range of potential climate changes; (3) assessing the vulnerability of LCRA to climate change impacts; (4) assessing system performance according to the uncertainty associated with climate change factors driving LCRA vulnerability; and (5) evaluating overall system risk and identifying areas in need of further analysis.

3.3. Policy Environment as a Constraint

A key constraining factor affecting any utility's ability to assess climate change vulnerability is the level of support of the local community about climate change, in particular utility ratepayers, boards of directors, and locally elected officials. In many circumstances, it is not feasible for a utility to engage in a "climate change" vulnerability study because there is no political support for the effort. Therefore, the policy environment directly impacts a utility's approach to climate change vulnerability assessment. For example, New York City Mayor Michael Bloomberg has taken climate change seriously during his tenure, providing political support for climate change adaptation efforts. As a result, NYCDEP has been able to do sophisticated analyses and make long-term investments to assess potential risks. Many water resource managers in less supportive political environments, however, do not have the political support needed to take on climate change directly or explicitly and must use historic or paleocli mate data rather than downscaled climate model data to justify operational changes or investments that require a rate increase. If support does not exist for climate studies, a water manager might elect to do a sensitivity analysis as a lower-profile way to understand the potential vulnerabilities of their water system to changes in climate conditions – climate change or climate variability.[7]

Vulnerability to changing environmental conditions is often incorporated into contemporary utility planning efforts even if those efforts are not billed as climate change vulnerability assessments. Indeed, during research for this report, we discovered a utility that reported taking sea level rise into account in their infrastructure design as far back as 1989, but without advertising the analysis as "climate change." In summary, the policy environment in which a utility operates may dictate what approaches to climate vulnerability assessment are acceptable, how such studies are conducted, what data sources are used, and other important aspects of vulnerability assessment design.

4. Sources of Climate Information

Regardless of how a climate change vulnerability assessment is framed, climate information that operates at the spatial and temporal scale of a utility's hydrology, planning, and/or operational models is almost always used in vulnerability assessments.[8] Most commonly, temperature and precipitation estimates at daily, weekly, or monthly timesteps are input into utility-specific hydrology, demand, operations, and/or management models to assess the effects of different climate conditions on water supply or wastewater services.[9]

Temperature and precipitation estimates can be derived from a number of sources, including the instrumental record, paleoclimate studies of variability beyond the historical record, literature reviews, and downscaling GCM projections to utility-specific watershed(s) to generate projected future climate conditions. These sources of climate information have different associated logistical and resource constraints and each is explored in more depth below. The key criterion for selecting a source of climate information is that the approach and the data can be justified, sometimes only to a technical audience, but increasingly to ratepayers, boards of directors, and elected officials as well.

Utilities need credible climate information in order to consider climate change impacts on their water systems. Historically, most utilities that address climate change issues have relied on partnerships with academic and research institutions to help them select appropriate climate data, however, many utilities are increasingly turning to private consulting companies to help them with this task. Small- and medium-sized utilities, however, often lack the financial resources needed to engage consultants to provide technical guidance on climate change vulnerability studies and their ability to partner with academic institutions is not clear. The scalability of the climate information approaches described below must be carefully considered for potential applicability beyond the relatively large and highly-resourced water utilities that have led the movement to consider climate change explicitly in their infrastructure and operational planning.

4.1. The Instrumental Record

The instrumental record is often used to define utility-specific models for demand, hydrology, operations, and management. These models can be simple statistical relationships or complex physical models of local hydrology that incorporate vegetation cover, aspect, slope angle, and soil type in addition to precipitation and temperature. Regardless, the instrumental record forms a fundamental basis for most utility vulnerability assessments in one form or another.

Utility vulnerability assessments have historically used the drought of record or flood of record to assess the vulnerability of a water system by applying the most extreme climate conditions on record to contemporary circumstances. In many cases, this type of analysis can be useful for identifying current vulnerabilities of a water system to current climate.[10] Although none of the utilities discussed in this report relied exclusively on this assessment methodology, many utilities still routinely rely on this method due to its familiarity and ease of use.[11] It is important to note that using only the instrumental record implicitly retains the assumption of climate stationarity – fluctuation within an unchanging envelope of variability – as observed typically within the last century or less.[12]

4.2. Paleoclimate Data

Some researchers have begun developing proxy records for precipitation and streamflow based on tree rings and geological evidence.[13] These researchers have found evidence of dry periods prior to the 20th century that were more persistent and perhaps more intense than what is observed in the instrumental record (e.g., Woodhouse and Lukas, 2006a). Because natural climate variability can often be quite large compared to changes projected from climate change, some utilities have used paleoclimate data as a complement to their climate vulnerability assessments.

For example, EBMUD used paleoclimate reconstructions of riverflow on the Sacramento River to examine the resilience of their drought planning sequence (Meko and Woodhouse, 2005). Their use of paleoclimate data was largely qualitative, but the analysis demonstrated EBMUD's drought planning sequence would likely function adequately under the drought conditions evidenced in the paleorecord. Denver Water has also used 375-year tree ring reconstructions of the Colorado and Platte River basins in conjunction with their water supply simulation model – PACSM – to estimate the frequency and severity of drought within Denver Water's collection system (Woodhouse and Lukas, 2006a). Boulder Utilities used a 300-year tree ring reconstruction of hydrology for similar purposes (Hydrosphere, 2003).

Tree-ring reconstructions hold much promise because of an excellent network of tree-ring data, especially (but not limited to) the western United States (TreeFlow, 2010). The use of paleoclimate reconstructions, however, is limited because tree-ring reconstructions are not available for all areas, they can be difficult to translate into the modeling environment for water management, good relationships with hydrology have only been established for the growing season – giving an incomplete record of annual hydrology, and the instrumental records necessary for high-quality calibration of tree-ring data can be difficult to obtain. For these reasons, and because of the specialized skill set necessary to conduct such studies, paleoclimate data may not be feasible for many water utilities (Garrick and Jacobs, 2005).

4.3. Literature Reviews

Literature reviews can generate information for projected future climate conditions for many utilities at low cost. The most common sources for initiating climate literature reviews are the Intergovernmental Panel on Climate Change Fourth Assessment Report (I PCC, 2007) and the U.S. Global Change Research Program's report *Global Climate Change Impacts in the United States* (Karl et al., 2009). The National Oceanic and Atmospheric Administration (NOAA) is currently developing a climate information web portal (http://www.climate.gov), and NOAA's Regional Integrated Sciences and Assessments (RISA) program offices often provide regional literature and data resources (http://www .climate. noaa.gov/cpo_pa/risa/).

With the dramatic increase of climate research across the United States, regional specification of GCM output can often be found in the scientific peer-reviewed literature or on web portals.

For example, EBMUD decided not to expend resources on developing its own downscaled climate projections because sufficient data could be obtained in works of the IPCC (2007), the U.S. National Research Council (NRC, 2002), and the U.S. Geological Survey (Dettinger, 2005). Using these data, EBMU D ran a sensitivity analysis on their water system using the following factors:

- Increased customer demand resulting from a 4°C (7.2°F) increase in air temperature between 1980 and 2040;
- Changes in the timing of runoff in the Mokelumne River corresponding to 2°, 3°, and 4°C (3.6°, 5.4°, and 7.2° F) increases in air temperature; and
- Reductions in Mokelumne River runoff corresponding to a 10% and 20% decrease in precipitation.

EBMUD combined these three factors to derive seven scenarios that represented an adequate range of conditions to analyze the potential sensitivity of the EBM UD water system to climate change.

4.4. Climate Projections

The development of watershed-specific projected future climate conditions is the most computationally and resource intensive approach to gathering climate information. Data obtained by this method were employed by many utilities, including the City of Boulder Utilities Division (Smith et al, 2009), Denver Water (Laurna Kaatz, personal communication, April 8, 2009), SPU (Palmer, 2007), MWRA (Stephen Estes-Smargiassi, personal communication, February 26, 2010), NYCDEP (2008), PWB (Palmer and Hahn, 2002), and LCRA (CH2M Hill, 2008). Climate projections consist of three main elements: the use of socioeconomic scenarios of GHG emissions, the selection of GCMs, and the region-specific downscaling of climate model output. Each of these elements is described in greater detail below. Also described below is the innovative combination of climate projections with other data sources for analytical purposes.

It is important to note that the complexity and resource intensity of making climate projections for water resources is being rapidly reduced. For example, the U.S. Bureau of Reclamation now provides a downscaled dataset of statistically downscaled monthly climate projections from 1950 to 2099 that consists of 112 climate projections based on 16 climate models and three GHG emissions scenarios downscaled to a spatial resolution of 1/8 degree across the coterminous United States (Maurer et al, 2007). The National Center for Atmospheric Research (NCAR) is also developing a dataset of GCMs combined with different regional climate models (RCMs) (UCA R, 2007). As these datasets and tools become more widespread, the time and resource threshold to engage in climate projections likely will be lowered. CH2M Hill"s assessment of the LCRA system used the Bureau of Reclamation dataset as the foundation for its analysis.

4.4.1. Climate scenarios

Utility climate scenarios generally consist of two distinct elements: GHG emissions scenarios and defined time periods. Probabilities have not been assigned to the emissions scenarios (they are all assumed equally likely to occur) and, as a result, probabilities have not been assigned to climate change scenarios.[14]

GHG emissions scenarios have been developed to identify plausible future global energy use, economic growth, land-use change, population growth, technological innovation, and other factors that could affect future emissions of GHGs. Most utilities in our analysis used GHG emissions scenarios defined in the 2000 IPCC Special Report on Emissions Scenarios (SRES; see Box 1).[15]

Box 1. IPCC SRES Scenarios

- The A1 storyline and scenario family describes a future world of very rapid economic growth, global population that peaks in mid-century and declines thereafter, and the rapid introduction of new and more efficient technologies. Major underlying themes are convergence among regions, capacity building, and increased cultural and social interactions, with a substantial reduction in regional differences in per capita income. The A1 scenario family develops into three groups that describe alternative directions of technological change in the energy system. The three A1 groups are distinguished by their technological emphasis: fossil intensive (A1FI), non-fossil energy sources (A1T), or a balance across all sources (A1B).
- The A2 storyline and scenario family describes a very heterogeneous world. The underlying theme is self-reliance and preservation of local identities. Fertility patterns across regions converge very slowly, which results in continuously increasing global population. Economic development is primarily regionally oriented and per capita economic growth and technological change are more fragmented and slower than in other storylines.
- The B1 storyline and scenario family describes a convergent world with the same global population that peaks in midcentury and declines thereafter, as in the A1 storyline, but with rapid changes in economic structures toward a service and information economy, with reductions in material intensity, and the introduction of clean and resource-efficient technologies.

The emphasis is on global solutions to economic, social, and environmental sustainability, including improved equity, but without additional climate initiatives.

- The B2 storyline and scenario family describes a world in which the emphasis is on local solutions to economic, social, and environmental sustainability. It is a world with continuously increasing global population at a rate lower than A2, intermediate levels of economic development, and less rapid and more diverse technological change than in the B1 and A1 storylines. While the scenario is also oriented toward environmental protection and social equity, it focuses on local and regional levels.

Source: Nakićenovic et al., 2000 (p. 4-5).

(i.e., A2) to ensure representation of a wide range of plausible futures.

In general, water utilities have selected two or three GHG emissions scenarios to provide substantially different future GHG emission levels. Using multiple GHG emissions scenarios avoids putting too much emphasis on any single projection of the future. For example, the SPU study used emissions scenarios A2 and B1; NYCDEP used emissions scenarios A2, A1B, and B1; and LCRA used emissions scenarios A2 and B1. While, by definition, these scenarios are equally plausible, water utilities have generally selected both a low emissions/environmentally friendly scenario (i.e., B1) and a high emissions/development-oriented scenario

Furthermore, nearly all utilities utilizing climate projections looked at two or three future time periods, often defined by relevance to their planning processes. Typically, projections were made for approximately 30 years for operations or strategic planning and approximately 50 years for infrastructure and capital improvement plans. For example, the SPU study used three time periods (2025, 2050, and 2075); Portland used two time periods (2025 and 2045); and LCRA used two time periods (2050 and 2080). Each time period generally represented an average of 10 of more years of climate data to smooth out year-to-year variation in model output.

4.4.2. Global climate models

After selecting appropriate scenarios, the water utilities generally identified a subset of GCMs to use in their study. GCMs are generally selected from the World Climate Research Programme's Coupled Model Intercomparison Project (CMIP3).[16] CMIP3 contains data contributed by 23 of the world's leading climate models (see Table 1).

This climate model output, however, can vary dramatically by region. Consequently, for this and other practical considerations, most water utilities select a subset of climate models to use in their study. Three common methods for selecting models are described below.

One method for selecting climate models is hindcasting, which refers to using climate models to project climate over years for which instrumental data are available. The ability of various models to replicate observed climate is often used as a means of selecting a subset of climate models for use in the study under the justification that those models more accurately model the climate of that watershed or region. For example, in the first phase of its work, the NYCDEP Climate Change Task Force selected three GCMs namely the National Center for Atmospheric Research (NCAR), Goddard Institute of Space Studies (GISS) and European Center Hamburg Model (ECHAM) (NYCDEP, 2008). In the second phase of work, NYCDEP evaluated the ~20

GCMs for various combinations of five meteorological variables (precipitation, average, maximum and minimum temperatures and wind speed), over four seasons using probability based skill scores. Since no single model performs best for all variables and seasons they arrived at the most objective way to choose a subset of models to identify the models with the highest mean skill scores (averaged across all variables).

Table 1. GCMs in the World Climate Research Programme's CMIP Phase Three

Model IPCC designation	First published	Sponsor	Country
BCC-CM1	2005	Beijing Climate Center	China
BCCR-BCM2.0	2005	Bjerknes Centre for Climate Research	Norway
CCSM3	2005	National Center for Atmospheric Research (NCAR)	USA
CGCM3.1(T47)	2005	Canadian Centre for Climate Modeling and Analysis	Canada
CGCM2.1(T63)	2005	Canadian Centre for Climate Modeling and Analysis	Canada
CNRM-CM3	2004	Meteo-France/Centre National de Recherches Météorologiques (CNRM)	France
CSIRO-MK3.0	2001	Commonwealth Scientific and Industrial Research Organisation Australia (CSIRO) Atmospheric Research	Australia
ECHAM5/MPI-OM	2005	Max Plank Institute for Meteorology	Germany
ECHO-G	1999	Meteorological Institute of the University of Bonn, Meteorological Research Institute of the Korea Meteorological Administration (KMA), and Model and Data Group	Germany/ Korea
Model IPCC designation	**First published**	**Sponsor**	**Country**
FGOALS-g1.0	2004	National Key Laboratory of Numerical Modeling for Atmospheric Sciences and Geophysical Fluid Dynamics (LA SG)/Institute of Atmospheric	China

		Physics	
GFDL-CM2.0	2005	U.S. Department of Commerce/NOAA Geophysical Fluid Dynamics Laboratory (GFDL)	USA
GFDL-CM2.1	2005	NOAA GFDL	USA
GISS-AOM	2004	National Aeronautics and Space Administration (NASA) Goddard Institute for Space Studies (GISS)	USA
GISS-EH	2004	NASA/GISS	USA
GISS-ER	2004	NASA/GISS	USA
INM-CM3.0	2004	Institute for Numerical Mathematics	Russia
IPSL-CM4	2005	Institut Pierre Simon Laplace	France
MIROC3.2 (high resolution)	2004	Center for Climate System Research (University of Tokyo), National Institute for Environmental Studies, and Frontier Research Center for Global Change (JAMSTEC)	Japan
MIROC3.2 (medium resolution)	2004	Center for Climate System Research (University of Tokyo), National Institute for Environmental Studies, and Frontier Research Center for Global Change (JAMSTEC)	Japan
MRI-CGCM2.3.2	2003	Meteorological Research Institute	Japan
PCM	1998	NCAR	USA
UKMO-HadCM3	1997	Hadley Centre for Climate Prediction and Research/Met Office	UK
UKMO-HadGEM1	2004	Hadley Centre for Climate Prediction and Research/Met Office	UK

A second method for selecting climate models is to purposefully select models that project a range of climate conditions. This was the strategy employed in the SPU study (Polebitski et al., 2007a). The three GCM-scenario combinations selected by the UW-CIG were the GISS Model ER/B 1 combination, the MPI ECHAM5/A2 combination, and the Institut Pierre Simon Laplace (IPSL) CM4/A2 combination. These models have performed well in other studies when replicating the temperature and precipitation trends of the Pacific Northwest during the 20th century (Mote et al., 2005). The MPI ECHAM5/A2 model represents a middle-of-the-road scenario with moderate warming and precipitation increase. The IPSL-CM4/A2 model scenario is significantly wetter and warmer, and the GISS-ER/B1 scenario is slightly drier and warmer.

The LCRA study (CH2M Hill, 2008) used a third method for selecting climate models that is effectively a hybrid of the two methods described above. Initially, six models were selected for further assessment based on expert judgment of the leading models for California selected by researchers at the Scripps Institute of Oceanography: NCAR CCSM3.0, NCAR PCM1, GFDL CM2.1, CNRM CM3, CCSR MICRO3.2 (medres), and MPI ECHAM5.[17] Projected changes in temperature and precipitation from all six models under two emissions scenarios (A2 and B1) and two time periods (2050 – averaged 2036 2065; and 2080 – averaged 2066 2099) were plotted onto a grid representing the 10th and 90th percentile of all 112 downscaled model simulations from the U.S. Bureau of Reclamation. Scenarios that plotted consistently outside of this range were discarded, resulting in the selection of two climate models for the LCRA study that were considered to provide a reasonable range of potential climate change effects in the Colorado River Basin.

4.4.3. Downscaling

Finally, all climate projections must be downscaled to the spatial and temporal resolution to be useful inputs into water system models. This is generally done using one of three downscaling methodologies: simple, statistical, or dynamic downscaling (see Box 2).

For example, the PWB study used simple downscaling. It started with relatively coarse spatial resolution model results – on the order of three degrees or approximately 300-km grids (Palmer and Hahn, 2002). UW downscaled the data to one-degree or approximately 100-km grids using the Symap algorithm. However, even at one degree, the model results were still coarse. Consequently, the climate change signal for each model and time period was calculated as the difference between average monthly temperature and precipitation from a control run simulating current climate and future climate model predictions. This yielded delta values by month for 2025 and 2045 that indicated the modeled change in temperature and precipitation. These deltas were then applied to an observed climate dataset used by UW as input into its Distributed Hydrology, Soil-Vegetation Model (DHSVM) to characterize the hydrology of PWB□s Bull Run watershed. This represents a form of simple downscaling because it used coarse GCM data and related them to the observed record to define a dataset at the spatial and temporal scale of the utility's models.

Box 2. Downscaling Methodologies

Simple downscaling involves adding projected changes in temperature and precipitation from GCMs to weather observations for a historical control period. This preserves observed spatial differences and locally observed seasonality of weather, and allows for the selection of different historic control periods for different purposes (e.g., a drought year or a year of heavy flooding). However, this method does not account for differences in climate change within a grid box.

Statistical downscaling is a method by which climate change can be projected for a particular location or small geographic area. Variables found in GCMs such as pressure are correlated with observations such as observed temperature or precipitation. The correlation uses GCMs" simulation of current climate. This correlation is used to project changes in the predicted variable(s) based on the estimated changes in GCMs. Statistical downscaling is computationally easier than dynamic downscaling, but does not allow for changes in the relationship between the predictor and predicted variables.

Dynamic downscaling uses RCMs to give higher resolution projections than GCMs. These models are like GCMs, but simulate only a portion of the globe and can therefore have much smaller grid boxes than GCMs and can incorporate sub-GCM-grid scale variables (e.g., mountain ranges, lakes). RCMs are "nested" within GCMs using "boundary conditions" from GCMs to drive them. RCMs typically have resolutions of 50 km or less. RCMs have the advantage of simulating dynamic relationships between climate variables, but require extensive computing power.

A key limitation of all three downscaling methods is that they depend on GCMs. Errors in GCMs are typically not corrected by downscaling techniques, although each technique will supply sub-GCM-grid scale variations in climate.

NYCDEP used a similar change factor approach, using the difference in daily GCM control and future scenarios to calculate change factors that were subsequently applied to local meteorological records used to drive watershed and reservoir models. In the first phase of NYCDEP climate change work, monthly change factors (additive or multiplicative) were calculated from pooled daily data associated with each month in each scenario. In the second phase of the work the change factor methodology was evaluated, and an improved method was developed that calculated 25 additive change factors for 25 equally spaced bins of the frequency distribution associated with each month"s pooled data. City of Boulder Utilities Division also used simple downscali ng, but combined GCM output with paleoclimate reconstructions.

The LCRA study used a different technique for downscaling. The selection of GCMs for the LCRA study ultimately led to eight datasets – two models (NCAR CCSM3.0 and GFDL CM2.1), two scenarios (A2 and B1), and two time periods (2050 – averaged 2036 2065; and 2080 – averaged 2066 2099). These eight datasets were statistically downscaled from the model grid scale to regional or watershed scales by retrieving the appropriate temporal-spatial data in the archived U.S. Bureau of Reclamation downscaled dataset (1/8 degree; 12-km grids).[18] A suite of data processing tools and geographic information systems were applied to these data to delineate the datasets for the State of Texas and the Colorado River Basin. Simulated future climate was analyzed by comparing the 2050 and 2080 periods with a reference period of 1970 1999 (CH2M Hill, 2008). This use of sophisticated statistical techniques to downscale GCM output to local watersheds is known as statistical downscaling.

The SPU study conducted its own method of statistical downscaling (Polebitski et al., 2007b). First, the SPU study did a bias correction of GCM simulation of hindcast climate using historic data. Those bias corrections were then applied to coarse (GCM grid scales) future climate data. The bias-corrected coarse grid was then related to a regional grid (1/8 degree) using scaling factors (additive factors for temperature and multiplicative factors for precipitation). The regional, bias-corrected future climate data are then related to specific weather station locations and again bias-corrected to yield a monthly time-series at the station level. These station level changes can be combined with historic data to yield scenarios of daily data that preserve historic climate variability.

For its statistical downscaling, the MWRA used a non-parametric method for generating the requisite daily weather sequences, specifically the K-nearest neighbor (K-NN) bootstrapping technique (Yates et al., 2003). This method samples the historical record in order to generate a large set of individual weather sequences that replicates the statistical characteristics of local weather but are consistent with the range of GCM-derived temperature and precipitation trends. Using probability density functions derived from 21 GCMs following the procedure outlined in Tebaldi et al. (2005), the K-NN algorithm was used to develop individual sequences of weather variables for the key weather station locations that are used by the system simulation model (WEAP) of MWRA's supply system.

The final downscaling methodology is dynamic downscaling. Dynamic downscaling uses RCMs to translate GCM output into higher-resolution regional climate data. Dynamic downscaling captures the effects of mesoscale features such as narrow mountain ranges, complex land/waterbody interactions, or variations in land use and land cover in a way that GCM currently cannot. Until recently, the computing power necessary for dynamic downscaling to provide useful data was prohibitive and the results were not high-resolution enough to improve upon statistical downscaling approaches. Current research into RCMs and dynamic downscaling shows promise. For example, UW-CIG is engaged in research to develop a fully functional RCM for the Pacific Northwest. NYCDEP also worked with Columbia University"s Center for Climate Systems Research (CCSR) and the City University of New York to establish a dynamic downscali ng strategy for future assessments.

4.4.4. Combining methods: Climate projections with paleoclimate data

The study conducted for the City of Boulder combined climate model output with paleoclimate data (Smith et al., 2009). To our knowledge, this is the first study to do so in the United States. The study matched paleoclimate reconstructions of streamflow in Boulder Creek with years during the observed climate record that had similar streamflow. The climate in the similar years was used as a proxy for the climate of the paleorecord. In other words, the reconstruction of streamflow in a paleo year (e.g., 1635) was then matched to a year in the observed record with a similar annual streamflow. A best match between reconstructed streamflow and years in the observed record did not always exist. Therefore, the technique used a random process to select years from the observed record with similar characteristics. The model was run many times to yield different combinations of the observed record and mimic the reconstruction. This introduces many combinations of reconstructions, which allows for more variability in results.

The Boulder study used a simple downscaling of GCM data. The changes in temperature and precipitation were added to or multiplied by temperature and precipitation, respectively, in the proxy record. This yielded a set of climate change scenarios that mimicked the year-to-year variability of the paleorecord, but with long-term average changes in climate from the GCMs imposed on the set.

The reconstruction of streamflow for Boulder Creek utilized a 437-year record from 1566 to 2002 (Woodhouse and Lukas, 2006b). The reconstruction technique estimates 65% (i.e., has an R2 of 0.65) of the observed streamflow variance.

The study used output from three GCMs under three emissions scenarios for the decades of the 2030s and 2070s. The selected emissions scenarios (A2, A1B, and B1) represented a wide range of future conditions. As with SPU, the GCMs were selected to capture a wide range of potential changes in regional climate. While all climate models project that the central Rocky Mountains and adjacent plains will become warmer, the models disagree as to whether precipitation will increase or decrease. To overcome this, the study used a relatively wet model (the Canadian Climate Model), a relatively dry model (GFDL Version 0), and a model roughly in the middle of the range of model output (GFDL Version 1).

5. Modeling Changes in Water Resources

Water utilities use a variety of models and analytical techniques for various purposes. As a general organizational distinction, it simplifies things to examine how a utility approaches water quantity and water demand separately.[19]

5.1. Water Quantity

Water utilities may use one or more water management model(s) for day-to-day operations and long-term water planning. Many, but not all, also use a hydrologic model to translate temperature and precipitation data into streamflow, reservoir storage, evaporative loss, and other variables input directly into water system management models. The specific constellation of models used by a utility often determines what climate information it needs to carry out a climate change vulnerability assessment.

PWB, for example, input temperature and precipitation data into the DHSVM to generate averaged annual hydrographs. D HSVM is a physically based hydrology model that characterized the entire Bull Run watershed into 150-m grid cells with grid-specific data on soil and vegetation type, soil depth, vegetation height, and surface elevation and slope. DHSVM represents the more detailed end of utility hydrology models that have been empirically characterized for a number of parameters in addition to temperature and precipitation.

PWB applied the UW-CIG developed climate altered streamflows through DHSVM and then input the data into PWB's Supply and Transmission Model (STM), an operational model of PWB's terminal water storage and groundwater resources. STM operates at a daily timestep and simulates the flow of water throughout the water transmission system using seasonally varying rule curves for the reservoirs, as well as modeling water releases for hydropower production and instream flows. Groundwater operations are coordinated with reservoir operations to enable the investigation of a variety of operating alternatives using variables such as length of drawdown period, amount of groundwater pumped during drawdown, minimum storage during drawdown, and water used during drawdown. As a part of a Water Research Foundation project, the PWB used the STM data sets to look at how the WEAP platform worked in comparison with the outputs of the STM, however, the WEAP model was not used for the initial climate change study done by UW-CIG. For purposes of this analysis it is important to note the use of multiple operational models for different purposes by a single utility.

In order to use SPU's systems model, climate change parameters needed to be converted to inflow. UW-CIG used the DHSVM hydrology model to produce climate-altered hydrologic datasets based on climate model output for use in the Conjunctive Use Evaluation (CUE). SPU then used the results from DHSVM in its CUE systems model – a weekly timestep simulation model of the Cedar and Tolt River systems – for water supply planning. CUE is used for calculating and evaluating the firm yield[20] and reliability of Seattle's water supply system and potential future water supply projects. CUE results indicated that yield decreased under all climate change scenarios for all time periods. SPU also ran several planning scenarios through CUE to determine whether available supply could be increased to compensate for anticipated shortfalls.

The use of river gauge data as input into CUE was a policy decision for SPU, because it also has the capability to use the Seattle Forecast Model (SEAFM), a proprietary hydrology model calibrated to the SPU watersheds and used in operational forecasting and operations planning. UW-CIG chose to use DHSVM instead of SEAFM (or its successor models) for its climate change vulnerability analysis because UW-CIG developed DHSVM and it was available to all watersheds in the broader UW-CIG led regional study in which SPU participated.

NYCDEP has a very complex modeling environment. It currently does analysis using the Variable Source Loading Function (VSL F) or the SWAT watershed models. These can in turn provide inputs to one of two reservoir models: a one-dimensional reservoir eutrophication model; or a two-dimensional reservoir turbidity transport model (CEQual W2). The OASIS reservoir system operation model requires measured or simulated water inputs to the reservoirs and simulates the storage and transfer of water between the 19 reservoirs comprising the NYC water supply system. The watershed models, take daily temperature and precipitation data to generate streamflow and evapotranspiration, as well as a number of water quality parameters. Outputs of the watershed models can be used to drive both the OASIS system model and reservoir water quality models. A knowledge of reservoir operations is an important determinate of reservoir water quality, and a needed input to the reservoir models. For future climate simulations, where historical operations are not known, OASIS simulation are needed to specify reservoir operation scenarios associated with a given climate scenario. All of the above models were in use at NYCDEP to evaluate the impacts of changes in watershed management, land-use, and reservoir operation policies. This integrated modeling system was easily adapted for evaluation of the effects of climate change, once credible future scenarios of the meteorological variables needed to drive the models were available.

5.2. Water Demand

Changes in climate can also have an important effect on water demand. Demand means different things to different utilities depending upon their customer base and the timeframe of their assessment. Several utilities evaluated in this report provided interesting examples of different approaches or concerns to project changing water demand. EBM U D was focused primarily on exterior water use by urban and suburban water customers, while SPU and PWB were focused on the balance between population growth and climate altered demand over time.

According to EBMUD, customer demands are projected to vary predominantly based on temperature change. While indoor water use is not expected to change significantly with global warming, outdoor water use could change dramatically. To account for this, 2040 customer demands were renormalized using a projected temperature increase of 2.15°C (3.87°F) between 2005 and 2040 with no change in precipitation. EBMUD□s analysis indicated that a 20% reduction in precipitation had little influence on overall customer demands compared to the projected temperature change. The demand model suggested a 3.6% increase in customer demands by the year 2040 (EBMUD, 2009).

The SPU study examined the effect of climate change on water demand using a dual approach of regression analysis and forecast modeling. First, SPU performed a regression analysis of peak season consumption for 1982 2007 using monthly consumption data, maximum temperature, and rainfall at SeaTac Airport for May through September. This relationship was assumed to hold in the future. SPU had already developed a demand forecasting model for its *2007 Water System Plan*, which forecasted non-climate altered demand change over time. Under this model, demand was forecasted to decrease below historic levels through 2050, but increase above historic levels by 2075. Applying the results of the regression analysis to these forecasts adjusts demand slightly upward due to climate change in 2025 and 2050. But in 2075, the climate- induced increase becomes more significant on top of the additional forecasted upswing from the non-climate altered demand model.

In 2002, PWB examined the effects of climate-altered streamflow on system operations from a demand perspective using their STM. The evaluation investigated the climate and population growth impact on demand and supply separately and then jointly to discern the discrete effects of climate change. The process for calculating the climate impact on demand was derived from the Joint Institute on the Study of Atmosphere and Oceans (JISAO) report, *Impacts of Climate Variability and Change in the Pacific Northwest* (JISAO-CIG, 1999). This study used seven featured years and the ECHAM4 climate scenario for detailed analysis. Growth in demand not considering changes in climate decreased the average minimum storage dramatically. This was exacerbated (approximately doubled) by the impacts of climate on hydrology and demand.

PWB also utilized their water demand econometric model to estimate near- and long-term water demand that established a relationship between total water demand and selected economic, demographic, seasonal, and day-to-day weather variables. This econometric model is used as a forecasting tool by projecting changes in the economic and demographic variables. In order to gauge the effects of climate change on demand, past weather years with patterns that approximate climate projections are selected to forecast future demands. For example, in projecting demands in 2050, PWB uses 1948 as the wettest, or best-case scenario (corresponding to a 3.67°F/2.04°C increase in average peak season temperature), 1980 as a middle-of-the-road scenario (corresponding to a 4.83°F/2.68°C increase in average peak season temperature), and 1991 as the driest, or worst-case scenario (corresponding to a 6.72°F/3.73°C increase in average peak season temperature).

6. SUMMARY

We investigated assessments of climate change vulnerability conducted by eight utilities to determine the emergent characteristics of water utility climate change vulnerability assessments (see Table 2). We found that two standard approaches were used for the risk assessments: the top-down modeling assessment and the bottom-up threshold analysis. While both methods produce important risk information, the selection of an assessment model is often determined by available fiscal and technical resources as well as the policy environment faced by the utility. These two approaches are not mutually exclusive, and are often run in series or in parallel to provide a robust vulnerability assessment. Notably, top-down assessments come in a variety of types, including scenario analyses, sensitivity analyses, and paleoclimate or historic analyses.

Sources of climate data used by the utilities included the instrumental record, paleoclimate data, literature reviews, and climate projections from GCMs. While information from each of these sources played an important role in one or more utility vulnerability assessments, climate projections offered the most computationally demanding approach. Climate projections required the selection of GHG emissions scenarios and defined time periods, the selection of a subset of GCMs (typically by hindcasting, selecting a broad range of outputs, or eliminating outliers), and downscaling the GCM data to the spatial and temporal scale necessary for input into utility hydrology, management, or operations models through simple, statistical, or dynamical downscaling techniques.

Table 2. Aspects of utility vulnerability assessment efforts

	Assessment approach	Climate model selection	GHG emissions scenarios	Downscaling	Time periods	System models	Runoff models
Boulder	Top-down (paleo)	Purposeful selection for range	A2, A1B, B1	Simple	2030, 2070	Boulder Creek Model	CLIRUN
Denver Water	Top-down (sensitivity,scenario in process)	Current study: Purposeful selection for range	Current study: A2, A1B, B1	Current study: statistical	In process study: 2040 and 2070	PACSM, ESP	Sacramento Soil Moisture coupled with Anderson Snow-17, WEAP
MWRA	Top-down	All models (Yates/NCAR project)	A1B	Statistical	Weekly projections through 2060	WEAP	abcd watershed model
NYCDEP	Top-down (scenario)	Probability based skill score approach	A2, A1B, B1	Delta change factor methodology by month and frequency distribution, statistical	2046-2065 and 2081-2100	OASIS, CEQual W2, 1-D Eutrophication	VSLF, SWAT
PWB	Top-down (sensitivity)	Purposeful selection for range	1% annual increase in CO_2[a]	Statistical	2025, 2045	STM, WEAP	DHSVM
SPU	Top-down (scenario)	Purposeful selection for range	2007 study: A2, B1, 2002 study: A2	Statistical	2007 study: 2000, 2025, 2050, 2075 2002 study: 2000, 2020, 2040	Conjunctive Use Evaluation (CUE)	DHSVM

Table 2. (Continued)

	Assessment approach	Climate model selection	GHG emissions scenarios	Downscaling	Timeperiods	System models	Runoffmodels
LCRA	Bottom-up (scenario)	Hindcast and purposeful selection for range	A2, B1	Statistical (Bureau of Reclamation dataset)	2050, 2080	WAMWRAP,	Variable Infiltration Capacity
EBMUD	Bottom-up (sensitivity)[b]	Conducted sensitivity analysis	N/A	N/A	N/A	EBMUDSim, WEAP	N/A

a. The 1% annual increase in CO2 scenario was often used prior to development of the SRES scenarios.

b. Because EBMUD chose to use a sensitivity analysis, the climate modeling aspects of this table are not applicable.

Finally, the climate information was run through utility-specific models to determine the effects of projected climate conditions on water supply and demand. We found that on the supply side, the models typically used included both hydrology and operational or management models. Demand modeling by utilities ranged from service-area-specific correlations between customer demand and temperature/precipitation to utility-specific demand models developed for long- range planning.

7. Recommendations for Further Study

This report describes a range of practices used by eight water utilities to evaluate the effects of climate change on water quantity and water demand using hydrological, management and operational models. Climate change is a complex issue and will require ongoing work to establish reliable practices for incorporating climate change into water utility decisions and planning. Additional questions to develop this line of inquiry further include:

- How are decision makers and rate payers responding to these types of analyses?
- Would the results vary if different methods were used by the same utility?
- How do the different methods compare in effectiveness, over time?
- What other decisions, utility models, etc. need climate vulnerability assessments?
- How are vulnerability assessments being conducted for understanding climate impacts on managing:
    - Water quality?
    - Intense precipitation and storms (stormwater, floods, wind)?
    - Sea level rise?

As other water resource managers gain more experience with climate vulnerability assessments, it would benefit the water resource management field to continue to document lessons learned in an effort to create, over time, a solid foundation of acceptable industry practices.

8. References

[1] Barsugli, J., Anderson, C. Smith, J.B. & Vogel, J.M. (2009). *Options for Improving Climate Modeling to Assist Water Utility Planning for Climate Change*. Final Report. Prepared for Water Utility Climate Alliance. December.

[2] CH2M Hill. (2008). *Climate Change Study: Report on Evaluation Methods and Climate Scenarios. Final Report.* Lower Colorado River Authority and San Antonio Water System. November.

[3] Dettinger, M.D. (2005). From climate-change spaghetti to climate-change distributions for 21st Century California. *San Francisco Estuary and Watershed Science*, *3*(1). Available: http://repositories.cdlib.org/ jmie/ sfews/vol3/iss1/art4.

[4] EBMUD. (2009). *WSMP 2040: Water Supply Management Program 2040 Plan*. East Bay Municipal Utility District. October.

[5] Freas, K., Bailey, B. Munevar, A. & Butler, S. (2008). Incorporating climate change in water planning. *Journal of the American Water Works Association*, *100*, 6. June.

[6] Garrick, D. & Jacobs, K. (2005). Tree-ring records inform water management decisions: Workshop strives for better integration of past with present. *Southwest Climate Outlook*. May.

[7] Hydrosphere. (2003). *Simulation of Hypothetical Climate Change Scenarios Using the Boulder Creek Watershed Model*. Hydrosphere Resource Consultants (now AMEC), Boulder, CO.

[8] IPCC. (2007). *Climate Change 2007: The Physical Science Basis. Contribution of Working Group I to the Fourth Assessment Report of the Intergovernmental Panel on Climate Change*, Solomon, S. Qin, D. Manning, M. Chen, Z. Marquis, M. Averyt, K.B. Tignor M. & Miller (eds.), H.L. Cambridge University Press, Cambridge, UK & New York.

[9] JISAO-CIG. (1999). *Impacts of Climate Variability and Change: Pacific Northwest*. A Report of the Pacific Northwest Regional Assessment Group for the US Global Change Research Program. Joint Institute for the Study of Atmosphere and Oceans – Climate Impacts Group. Office of Global Programs at the National Oceanographic and Atmospheric Administration. Seattle, WA. November.

[10] Karl, T.R. Melillo, J.M. & Peterson, T.C. (2009). *Global Climate Change Impacts in the United States*. Cambridge University Press, New York.

[11] Major, D.C. Horton, R. & Rosenzweig, C. (2007). *Integrated Modeling Project for Water Quantity and Quality: Guidelines for Project Final Report.* Climate Impact Assessment of Environmental Infrastructure Systems. New York City Department of Environmental Protection. June 29.

[12] Maurer, E.P. Brekke, L. Pruitt, T. & Duffy, P.B. (2007). Fine-resolution climate projections enhance regional climate change impact studies. *Eos Trans. AGU, 88*(47), 504.

[13] Means III, E.G. Laugier, M.C. & Daw, J.A. (2010). *Evaluating Decision Support Planning Methods for Incorporating Climate Change Uncertainties into Water Planning. Final Report.* Prepared for Water Utility Climate Alliance. January.

[14] Meko, D.M. & Woodhouse, C.A. (2005). Tree-ring footprint of joint hydrologic drought in Sacramento and Upper Colorado River basins, western USA. *Journal of Hydrology, 308*, 196-213.

[15] Miller, K. & Yates, D. (2006). *Climate Change and Water Resources: A Primer for Municipal Water Providers.* Jointly sponsored by Awwa Research Foundation, Denver, CO and University Corporation for Atmospheric Research, Boulder, CO. Awwa Research Foundation, American Water Works Association, and IWA Publishing.

[16] Milly, P.C.D. Betancourt, J. Falkenmark, M. Hirsch, R.M. Kundzewicz, Z.W. Lettenmaier, D.P. & Stouffer, R.J. (2008). Stationarity is dead: Whither water management? *Science, 319*, 573-574.

[17] Moss, R.H. Edmonds, J.A. Hibbard, K.A. Manning, M.R. Rose, S.K. van Vuuren, D.P. Carter, T.R. Emori, S. Kainuma, M. Kram, T. Meehl, G.A. Mitchell, J.F.B. Nakicenovic, N. Riahi, K. Smith, S.J. Stouffer, R.J. Thomson, A.M. Weyant, J.P. & Wilbanks, T.J. (2010). The next generation of scenarios for climate change research and assessment. *Nature, 463*, 747-756.

[18] Mote, P.W. Salathe, E. & Peacock, C. (2005*). Scenarios of Future Climate for the Pacific Northwest.* University of Washington: Climate Impacts Group.

[19] Nakićenovic, N. Alcamo, J. Davis, G. de Vries, B. Fenhann, J. Gaffin, S. Gregory, K. Grubler, A. Jung, T.Y. Kram, T. La Rovere, E.L. Michaelis, L. Mori, S. Morita, T. Pepper, W. Pitcher, H. Price, L. Raihi, K. Roehrl, A. Rogner, H.-H. Sankovski, A. Schlesinger, M. Shukla, P. Smith, S. Swart, R. van Rooijen, S. Victor, N. & Dadi, Z. (2000). *Emissions Scenarios. A Special Report of Working Group III of the*

Intergovernmental Panel on Climate Change. Cambridge University Press, New York.

[20] UCAR. (2007). North American Regional Climate Change Assessment Program. Available: http://www.narccap.ucar.edu/. Accessed May 28, 2010.

[21] NRC. (2002). *Abrupt Climate Change: Inevitable Surprises*. National Academies Press: Washington, D.C. Available: http://www.nap.edu/openbook.php?isbn = 0309074347.

[22] NYCDEP. (2008). *The New York City Department of Environmental Protection Climate Change Program Assessment and Action Plan: Report 1*. May.

[23] Palmer, R.N. (2007). *Final Report of the Climate Change Technical Committee*. A report prepared by the Climate Change Technical Subcommittee of the Regional Water Supply Planning Process, Seattle, WA.

[24] Palmer, R.N. & Hahn, M. (2002). *The Impacts of Climate Change on Portland's Water Supply: An Investigation of Potential Hydrologic and Management Impacts on the Bull Run System*. Portland Water Bureau. January.

[25] Polebitski, A. Traynham, L. & Palmer, R.N. (2007a). *Technical Memorandum #4: Approach for Developing Climate Impacted Meteorological Data and its Quality Assurance/Quality Control*. A report prepared by the Climate Change Technical Subcommittee of the Regional Water Supply Planning Process, Seattle, WA.

[26] Polebitski, A. Wiley, M.W. & Palmer R.N. (2007b). *Technical Memorandum #2: Methodology for Downscaling Meteorological Data for Evaluating Climate Change*. A report prepared by the Climate Change Technical Subcommittee of the Regional Water Supply Planning Process, Seattle, WA.

[27] Smith, J.B. Strzepek, K. Rozaklis, L. Ellinghouse, C. & Hallett, K.C. (2009). *The Potential Consequences of Climate Change for Boulder Colorado's Water Supplies*. Prepared for NOAA's Climate Program Office. February.

[28] Strange, E.M. Lane, D.R. & Herrick, C.N. (2009). Utility Guidance for Mitigating Catastrophic Vegetation Change in Watersheds. Project #4009. Water Research Foundation, Denver, CO.

[29] Stratus Consulting & MWH Global. (2009). *Implications of Climate Change for Adaptation by Wastewater and Stormwater Agencies*. Water Environment Research Foundation Report No. CC2R08.

[30] Tebaldi, C. Arblaster, J.M. Hayhoe, K. & Meehl, G.M. (2006). Going to extremes: An intercomparison of model-simulated historical and future change extremes. *Climatic Change*, *79*, 185- 211.

[31] Tebaldi, C. Smith, R.L. Nychka, D. & Mearns, L.O. (2005). Quantifying uncertainty in projection of regional climate change: A Bayesian approach to analysis of multi-model ensembles. *Journal of Climate*, *18*, 1524-1540.

[32] TreeFlow. (2010). Streamflow reconstructions from treerings website. Available: http://treeflow.info/. Accessed March 30, 2010.

[33] U.S. EPA. (2009). *Proceedings of the First National Expert and Stakeholder Workshop on Water Infrastructure Sustainability and Adaptation to Climate Change*. Publication No. EPA 600/R-09/010. Available: http://www.epa.gov/nrmrl/wswrd/wqm/wrap/workshop.html.

[34] Woodhouse, C.A. & Lukas, J.J. (2006a). Multi-century tree-ring reconstructions of Colorado streamflow for water resource planning. *Climatic Change*. DOI: 10.1007/s1 0584-006-9055-0.

[35] Woodhouse, C. & Lukas, J. (2006b). Streamflow Reconstructions for Boulder Creek, South Boulder Creek, and the Colorado River. Report to City of Boulder and Hydrosphere Resource Consultants. AMEC Consulting, Boulder, CO.

[36] WRF. (2010a). Analysis of Changes in Water Use Under Regional Climate Change Scenarios. Project #4263. Project in progress; completion year 2013. Water Research Foundation, Denver, CO.

[37] WRF. (2010b). Estimating Salinity Effects Due to Climate Change on the Georgia and South Carolina Coasts. Project #4285. Project Description; completion year 2012. Water Research Foundation, Denver, CO.

[38] WRF. (2010c). Evaluating Effects of Climate Change on Water Utility Planning Criteria and Design Standards. Project #4154. Project Description and Abstract; completion year 2010. Water Research Foundation, Denver, CO.

[39] WRF. (2010d). Incorporating Climate Change Information in Water Utility Planning: A Collaborative, Decision Analytic Approach. Project #3132. Project Description and Abstract; completion year 2010. Water Research Foundation, Denver, CO.

[40] WRF. (2010e). Project Snapshot: Joint Front Range Climate Change Vulnerability Study #4205. Available: http://www.waterresearch foundation.org/research/TopicsAndProjects/projectSnapshot.aspx?pn = 4205. Accessed March 30, 2010.

[41] WRF. (2010f). Water Utilities and Climate Change: A Research Workshop on Effective System Adaptation. Project #4228. Project Description; completion year 2010. Water Research Foundation, Denver, CO.

[42] Yates, D. Gangopadhyay, S. Rajagopalan, B. & Strzepek, K. (2003). A technique for generating regional climate scenarios using a nearest neighbor algorithm, *Water Resources Research*, *39*(7), 1199.

End Notes

[1] A forthcoming report from the Water Research Foundation (WRF, 2010d) offers case study information relevant to this report, but the final draft was not publicly available for inclusion in this report.

[2] Furthermore, most utilities take land use changes, population projections, and economic development into account in infrastructure design and operational planning. Climate, however, often gets lumped in with hydrological considerations and does not rise to the level of an independent planning consideration.

[3] This conclusion was also reached in a study of the Water Utility Climate Alliance utilities (Barsugli et al., 2009).

[4] The term "downscaling" has come to mean the use of higher resolution regional climate models (RCMs; dynamic downscaling) or complex statistical approaches (statistical downscaling), but in this case can also refer to simple downscaling, which typically involves adding the temperature increase in a GCM grid to observed temperatures for stations in the grid box and multiplying the percentage change in precipitation by the observed precipitation record.

[5] Sometimes sensitivity analyses use arbitrary increments, but they can also be informed by the scientific literature. Sensitivity analyses can use different changes in temperature and precipitation by season, but this approach was not used in the vulnerability analyses investigated for this report.

[6] The lack of evidence for sole reliance on these types of analyses in the cases investigated for this report is largely a product of the case selection methodology, which focused on utilities with a published record of sophisticated vulnerability work.

[7] Water utilities might also select these methods for other reasons, such as a lack of funding or because uncertainty in the projections does not yield enough value added to justify the effort of an intensive climate change vulnerability assessment.

[8] Note that climate information includes more than GCM output, including historic weather station data, stream gauge data, paleocli mate reconstructions, and more. However, when GCM output is used, it typically has to be downscaled or assumptions made to reduce the temporal and spatial resolution to be usable in utility hydrology or operational models.

[9] Climate models often provide monthly average changes that must then be combined with the observed record at a daily ti mestep to generate projected daily changes in climate.

[10] With a changing climate, there should be more caution about interpreting the historic record. What may have been a 100-year drought or flood calculated from the historic record may shift significantly with projected climate change, or even by adding the paleorecord to historic data.

[11] For example, both Contra Costa Water District and Santa Clara Valley Water District in California make conservative assumptions about "worst-case scenarios" for drought planning instead of engaging in climate change-specific studies as described in this report.

[12] The exception to this would be to identify trends in the historic record and extrapolate them into the future. None of the utilities surveyed carried out such an approach.

[13] The scientific field of dendrohydrology involves using tree-ring qualities, such as the width of the annual growth rings, to estimate pre-instrumental streamflow of a specific river.

[14] NCAR has developed probabilities of regional changes in temperature and precipitation based on analysis of GCM output (Tebaldi et al., 2005, 2006), but these probabilities were not used by the utilities examined in this study.

[15] IPCC is now moving toward a set of scenarios based on radiative forcing and not tied to specific socioeconomic scenarios (Moss et al., 2010).

[16] Note that CM IP is in the process of updating model outputs in anticipation of the Fifth Assessment Report of the IPCC. The next iteration of CMIP will be called CMIP5 – skipping "CMIP4" – in order to align the numbering with the IPCC reports.

[17] NCAR CCSM3.0 and GFDL CM2.1 were ultimately chosen for use in the LCRA study.

[18] The Bureau of Reclamation developed an archive of statistically downscaled and bias-corrected climate projections for the contiguous 48 states at 1/8 degree resolution, which is what was used in the LCRA study (see http://gdo-dcp.ucllnl.org/ downscaled_ cmip3_projections/ dcpInterface.html). It should be noted that it is possible to statistically downscale GCM projections using a variety of techniques independent of the U.S. Bureau of Reclamation dataset. The SPU study provides an example of a different technique.

[19] Some utilities in this study, such as NYCDEP, were also concerned about climate change effects on water quality. However, generally speaking, water quality involves a greater variety of models (e.g., reservoir specific eutrophication models) and/or micro-climate processes that often occur at a higher resolution than can be captured by current GCMs or RCMs (e.g., intense precipitation events). The scope of this report did not allow us to fully explore water quality modeling.

[20] Firm yield is based on the 98% reliability standard.

In: Water Utility Approaches and Responses... ISBN: 978-1- 61942-784-6
Editors:J. A. Williams and C. E. Jones:

Chapter 2

CLIMATE CHANGE VULNERABILITY ASSESSMENTS: FOUR CASE STUDIES OF WATER UTILITY PRACTICES*

U.S. Environmental Protection Agency

FOREWORD

Global climate change can have a range of potentially adverse effects on water resources. EPA's Global Change Research Program (GCRP), within the Office of Research and Development (ORD), conducts research to support the development of sustainable solutions for protecting human health and the environment from the effects of air pollution and climate change while sustainably meeting the demands of a growing population and economy. The focus on developing sustainable solutions requires innovative, systems-oriented science that engages end- users from problem-formulation through product application. GCRP research activities addressing the implications of climate change on water quality supports EPA's mission to protect human health and the environment, and is consistent with the goals outlined in the EPA Office of Water's *National Water Program Strategy: Response to Climate Change.*

* This is an edited, reformatted and augmented version of the United States Environmental Protection Agency, 600/R-10/077F, dated March 2011.

This report was developed in partnership with the EPA Office of Water (EPA OW). The report is a companion to a report released by EPA OW in 2010 titled *Climate Change Vulnerability Assessments: A Review of Water Utility Practices.* The 2010 report from the Office of Water describes the range of different approaches being applied by eight U.S. water utilities to assess their vulnerability to climate change. The current report expands on this discussion by presenting a series of case studies that describe in more detail the specific issues, analyses, and actions taken by four U.S. water utilities to assess and respond to climate change. The issue of climate change is complex and will challenge utilities as they strive to meet water quality goals. We hope that the case studies presented in this report will inform, inspire, or otherwise support the efforts of other utilities and water managers to better understand and respond to the challenge of climate change.

We want to thank the authors, reviewers, and entire project team for their effort in preparing this report. Success in producing this report has depended first and foremost on the dedication and enthusiasm of this team. We also want to acknowledge the collaboration of EPA Office of Research and Development and Office of Water in this effort, as well as the cooperation of the utilities examined. While much remains uncertain concerning future climate change, it is clear that our best path forward is through continued research, partnership and collaboration to meet our common goals.

Andy Miller
Acting National Program Director for Global Change Research
Office of Research and Development
U.S. Environmental Protection Agency

Karen Metchis
Senior Policy Advisor for Climate Change
Office of Water
U.S. Environmental Protection Agency

ACKNOWLEDGMENTS

This report could not have been possible without the generous help and support of staff from each of the utilities featured in the report: Alan Cohn, New York City Department of Environmental Protection; Dennis Diemer, East

Bay Municipal Utility District; Paul Fleming and Joan Kersnar, Seattle Public Utilities; and Rebecca West, Spartanburg Water. The authors also wish to thank Karen Metchis, Y. Jeffrey Yang, and Meredith Warren at U.S. EPA, and the entire project team at Abt Associates, Stratus Consulting, and U.S. EPA for their help and many thoughtful comments and suggestions that contributed to this report.

EXECUTIVE SUMMARY

Concern about the potential effects of climate change on the quantity, quality, timing, and demand for water is growing. The implications of climate change for long-lived, capital-intensive water infrastructure is a particular concern. In 2009, the U.S. Environmental Protection Agency sponsored the *First National Expert and Stakeholder Workshop on Water Infrastructure Sustainability and Adaptation to Climate Change* (U.S. EPA, 2009a). This workshop highlighted a need for improved information and tools to help water utilities better understand and manage the risk of future impacts due to climate change. Many utilities are already addressing this challenge. One finding of this workshop was that compiling case studies of current water utility activities related to climate change adaptation would be useful to help those in the water sector learn from each other.

This report presents case studies describing the approaches taken by four water utilities in the United States to assess their vulnerability to climate change. The report is not intended to be a comprehensive listing of assessment approaches or utilities conducting vulnerability assessments. Nor does this report represent the full range of potential approaches for assessing and managing climate risk. Rather, the purpose of this report is to illustrate a range of approaches that selected water utilities have taken to understand and respond to climate risk.

Water resources decision making is a complex, multi-factorial process, and determining the impact of a single factor, such as climate change, on decision making is a challenging question. In this report we present the facts from which utility managers, resource planners, climate scientists, and others can draw conclusions and lessons learned. This report is descriptive by design. It does not evaluate the approaches used. Furthermore, because this report is empirical in nature, it might not describe all topics that water utilities need to consider to assess their climate risk.

The following four utilities are featured as case studies in this report:

- East Bay Municipal Utility District (Contra Costa and Alameda Counties, California)
- New York City Department of Environmental Protection (New York, New York)
- Seattle Public Utilities (Seattle, Washington)
- Spartanburg Water (Spartanburg, South Carolina).

The selected utilities differ in terms of their geographic location, size, and the types of impacts they could face from climate change.

East Bay Municipal Utility District

East Bay Municipal Utility District (EBMUD) used an elaborate policy analysis when designing its Water Supply Management Program (WSMP) 2040 (WSMP 2040; EBMUD, 2009b). The objective of the WSMP 2040 was to identify and recommend a portfolio of projects for meeting dry-year water needs through 2040.[1] The WSMP 2040 process consisted of identifying potential adaptations, bundling them into 14 different portfolios, screening those portfolios based on historic hydrology, and then modeling 5 portfolios under climate change scenarios. EBMUD applied a "bottom-up" approach for the analyses by identifying climate factors most likely to affect the system's reliability and testing that reliability to changes in those factors that are projected to occur by 2040 (e.g., a 4°F increase in average daily temperatures between 1980 and 2040 or a 20% decrease in precipitation) (EBMUD, 2009a). EBMUD's analyses reaffirmed the need for a strategy that is flexible and is adaptable to further changes in observed climate and to refinements in climate change projections (EBMUD, 2009b).

New York City Department of Environmental Protection

To analyze vulnerability, the New York City Department of Environmental Protection (DEP) examined potential impacts of climate change on the availability of water, turbidity, and eutrophication. The vulnerability analyses identified several potential challenges to New York City's water supplies and quality, including increased demand, reduced inflows during the spring thaw season, and increased risk of nutrient loadings and eutrophication. Additionally, precipitation changes, sea level rise, and consequent increased salinity levels in the Hudson River could pose challenges to the City's drainage and wastewater treatment systems. DEP has identified a wide array of initiatives to reduce risks from these potential outcomes. Initiatives include developing a model-based reservoir operation support tool that will allow reservoir operations to be tailored to future climate conditions, relying more on the soon-to-be-filtered Croton water supply during turbidity events, building cost-effective grey infrastructure, making use of natural features such as the wetlands in Staten Island's Bluebelt, and promoting water conservation. DEP's extensive vulnerability studies have leveraged momentum for climate change considerations in both strategic and capital planning. For instance, DEP promotes the benefits of green infrastructure for adapting to climate change impacts (e.g., increased heavy precipitation events and intensifying of the urban heat island effect), as part of its broad, citywide effort to better manage stormwater.

Seattle Public Utilities

Seattle Public Utilities (SPU) has worked closely with the Climate Impacts Group (CIG) at the University of Washington (UW) since 2002 on two studies to assess climate change impacts. In the most recent study, UW-CIG selected global climate models to capture a range of conditions, downscaled them statistically, and ran the outputs through a hydrology model. These results were used in SPU's water supply planning model to examine the effect of climate change on SPU system performance. SPU also used the downscaled data to project changes in demand for water. All climate change scenarios modeled resulted in an estimated decrease in water supplies. The most direct use of the vulnerability assessment by SPU was for water planners to test the effect of different operational assumptions on water supply availability. SPU also identified far-reaching adaptations to use in future decades if demand exceeds water supplies because of either population growth or climate change.

Spartanburg Water

Spartanburg Water is an example of a relatively small utility that did not conduct quantitative vulnerability assessments (e.g., model-based assessments), but was able to use information on climate change together with recent extreme climate events to consider climate change in management decisions. South Carolina has experienced several extreme droughts and hurricanes in recent years and anticipates that climate change will exacerbate these extreme events. With lower low flows in receiving streams during droughts, wastewater treatment plants could be required to upgrade their technology to reduce discharge loads. Also, more intense precipitation could result in greater pollutant loadings from runoff to the receiving streams. In response to these concerns and to plan for projected population increases, Spartanburg Water made several changes in its infrastructure and operations. Recent concerns about water shortages led Spartanburg Water to assert its rights to limit water withdrawals from the reservoir for lawn irrigation during droughts. The utility also launched an aggressive water conservation program and, upon the installation of new pipes, decided to keep the old pipes in place for additional capacity to handle surges from stormwater. These adaptations are consistent with Spartanburg Water's experience with recent extreme events and concerns about population growth and climate change.

Observations across the Case Studies

The following summary observations can be made based on these case studies regarding the conduct and use of climate change vulnerability assessments to support adaptation:

- *Conducting climate change vulnerability assessments appears to have increased awareness of climate change risks, informed decision making, and supported adaptation at the utilities featured in this report.* One theme emerging from the case studies is the need to consider climate change in a holistic context, taking into account all factors affecting system performance.
- *Utilities have worked with climate scientists and modelers to obtain data and gain insight into how climate science can be used to inform their decision making.* SPU collaborated with the UW-CIG, DEP collaborated with Columbia University and the City University of New York, and EBMUD used an analysis conducted by the State of California and the California Climate Change Center. In contrast, Spartanburg relied on information gathered from briefings and staff contact with other utilities through participation in the Water Environment Federation and the American Water Works Association

but did not formally collaborate with the climate change research community to develop region-specific data on their climate change risks.

- *Uncertainties in vulnerability assessments or climate science need not delay adaptation action.* All four utilities described in this report have taken action to reduce their climate risk despite significant, remaining uncertainties regarding the potential impacts of climate change. Often, these actions also address other concerns, such as managing limited supplies using water conservation or water reuse. Nonetheless, climate change vulnerability analyses can help inform and prioritize among management decisions.
- *The large utilities used a wide array of climate change scenarios to capture some of the uncertainty about future climate change.* EBMUD conducted a "bottom-up" approach by performing sensitivity analyses to improve its understanding about how climate change could affect particular elements of its water resource system. SPU and DEP conducted what are often referred to as "top-down" approaches driven by climate change scenarios and models.[2]
- *Vulnerability analyses to date have focused mainly on water supply and demand.* All four utilities focused their climate change impacts and vulnerability work principally on water supply and demand. Although the utilities expressed concern about the effects of climate change on water quality, urban drainage, wastewater, and other aspects of their systems, these areas have not yet received the same level of attention as water supply.
- *The utilities used system-specific models to understand and manage potential climate impacts on their systems.* All studies except Spartanburg Water used their system hydrology, operations, or planning models to evaluate the effects of potential climate change on their systems. The models were used to assess whether operational changes would be sufficient to cope with the effects of climate change, or whether system changes, such as adding supplies or further reducing demand, also were necessary. These models did not always align with the output of climate models, necessitating tailored use of climate projections for each utility. Spartanburg Water used existing system models to understand their system's behavior and qualitatively determine expected climate change impacts.
- *The case studies demonstrate that a variety of methods can be used to understand vulnerability and analyze adaptation options.* Scenario

analyses, sensitivity analyses, state-of-the-science literature reviews, and peer information sharing were used in different combinations in the four case studies to understand the potential impacts of climate change. The sensitivity analysis EBMUD performed on their existing system model and the literature review and information gathering from peers Spartanburg Water performed demonstrate two paths that utilities lacking the financial and staff resources to support detailed modeling studies can take to assess their vulnerability to climate change.

- *Utilities expressed an interest in having their needs reflected in future research.* Utilities specifically requested higher resolution climate change projections for the spatial and temporal scales at which they operate, probabilities associated with projected changes, and guidance on appropriate climate change parameters and scenarios to consider and plan for in their regions. One recommendation was to establish a central repository of data to support climate change and adaptation analysis.
- *The results of vulnerability assessments by the four utilities were used in different ways to inform and support adaptation.* SPU responded directly to the results of the vulnerability analysis by evaluating the impact of conservative operational assumptions on reservoir management. The other utilities used their vulnerability assessments to increase knowledge about their climate change risks, integrate information on these risks into decision making, and provide support for adaptation measures.

1. Introduction

Concern about the potential effects of climate change on the quantity, quality, timing, and demand for water[3] is growing. In particular, decisions about water infrastructure have long-term implications because the infrastructure built today likely will be in place for decades. In 1997, the American Water Works Association (AWWA) issued a statement expressing the need for water utilities to begin planning for consequences of climate change (AWWA, 1997). In 2004, AWWA teamed with the National Center for Atmospheric Research to publish guidance for municipal utilities on how to address climate change (Miller and Yates, 2006). Three years later, eight major municipal water utilities formed the Water Utility Climate Alliance

(WUCA) to "provide leadership and collaboration on climate change issues affecting the country's water agencies" (WUCA, 2010).

Vulnerability to climate change, as defined by the Intergovernmental Panel on Climate Change (IPCC), refers to the exposure, sensitivity, and adaptive capacity of systems to climate change (Smit et al., 2001). Exposure consists of the type of change a system experiences. A coastal city might be exposed to a 3-foot sea level rise, while an inland city would not. Sensitivity is the effect that climate change can have on a system assuming no planned adaptation. For example, climate change is projected to reduce the growth of many crops but increase the growth of others. The sensitivity of these crops to climate change differs. Adaptive capacity refers to the potential or ability of a system to adapt to the effects of climate change (Smit et al., 2001). The adaptive capacity of a system is important, for example, in distinguishing the vulnerability of wealthy and poor societies or human systems versus ecosystems. Wealthier societies, in general, have greater adaptive capacity and, thus, on average, are considered less vulnerable to climate change than poorer societies (Parry et al., 2007).

Several water utilities have begun to assess the potential vulnerability of their systems to climate change. Many are considering whether their infrastructures or operations should be changed now or in the future to adapt to climate change. In 2009, the U.S. Environmental Protection Agency (EPA) sponsored the *First National Expert and Stakeholder Workshop on Water Infrastructure Sustainability and Adaptation to Climate Change* (U.S. EPA, 2009a). This workshop highlighted a need for improved information and tools to help water utilities better understand and manage the risk of future impacts due to climate change. One finding of this workshop was that compiling case studies of current water utility activities related to climate change adaptation would help others in the water sector learn from each other.

This report presents case studies describing the approaches taken by four water utilities in the United States to assess their vulnerability to climate change. The report is not intended to be a comprehensive listing of assessment approaches or utilities conducting vulnerability assessments (for more detail on these topics, see the companion report, U.S. EPA, 2010). Rather, its purpose is to illustrate a range of issues and current approaches taken by selected utilities that are leaders in climate adaptation to understand and respond to climate risk. Climate change and its effect on water resources is complex and will require ongoing attention and study. We hope the information gleaned from these case studies will be of use to water utilities and other members of the water resources community in illustrating a range of vulnerability studies being applied to guide adaptation decision making. This report is also intended to help identify the types of technical assistance needed to support such assessments.

A companion report is available, *Climate Change Vulnerability Assessments: A Review of Water Utility Practices* (U.S. EPA, 2010). The purpose of that report is to identify and categorize the models and techniques that eight water utilities are using to understand their vulnerability to climate change. The current report provides a more detailed examination of the approaches taken by three of the eight utilities discussed in U.S. EPA (2010). The current report also includes discussion of one utility not included in U.S. EPA (2010).

1.1. Selection of Case Studies

Many water utilities are active in developing climate adaptation strategies and could have been included here. Limiting the scope of this report to just four utilities, however, was necessary for practical reasons. The four utilities featured in this report are (Figure 1)

- East Bay Municipal Utility District (EBMUD) in Contra Costa and Alameda Counties, California
- New York City Department of Environmental Protection (DEP) in New York, New York
- Seattle Public Utilities (SPU) in Seattle, Washington
- Spartanburg Water in Spartanburg, South Carolina.

Figure 1. Location of water utility case studies.

The selected utilities differ in terms of their geographic location, size, and the types of impacts they could face from climate change (Table 1). Three of the four serve more than a million people. The smaller Spartanburg Water was selected because of its size and because it took a qualitative approach to understanding its vulnerability to climate change. The western utilities are mainly concerned about potential changes in the timing of and reductions in runoff, while the eastern utilities are concerned about changes in extreme events and consequences of these events for water quantity and quality, and the performance of their systems.

Each selected utility has examined or is examining the vulnerability of its system to climate change. The methods used span a range from detailed, quantitative analyses to a more qualitative approach for examining climate change and lessons learned from recent extreme events. All four utilities have also considered or made changes to planning, operations, or infrastructure that, if not driven by the results of their analyses, are at least consistent with adapting to climate change. These four case studies are not necessarily representative of how all utilities are considering climate change. Nor do they represent the full range of potential approaches for assessing and managing climate risk. They do, however, illustrate and provide insight into how information on vulnerability to climate change is being developed and used.

Table 1. Key attributes of water utility case studies

Utility	Location	Population served	Key climate change risks
East Bay Municipal Utility District (EBMUD)	Alameda and Contra Costa Counties, California	1.3 million	• Change in timing of runoff • Reduction in water supply • Sea level rise
Department of Environmental Protection (DEP)	New York, New York	9.2 million	• Increases in turbidity, eutrophication, and combined sewer overflows • Sea level rise
Seattle Public Utilities (SPU)	Seattle, Washington	1.4 million	• Change in timing of runoff • Reduction in water supply • Increases in flood risks and combined sewer overflows
Spartanburg Water	Spartanburg, South Carolina	180,000	• Increases in drought and coastal storms

1.2. Data Collection

The information presented in this report was collected from publically available documents and interviews with utility staff. Specifically, the report focuses on

- Background of the utility—e.g., location, size of utility
- Description of the utility, including the water supply (which includes provision of drinking water) and wastewater system
- Climate change projections and why the utility was interested in vulnerability to climate change
- Approach for conducting vulnerability assessment, including scenarios, assessment methods, and results
- Discussion of application of vulnerability assessment information.

Individual utility case studies are presented in the following four chapters. To the extent possible, the level of detail for the case studies is consistent. The final chapter of this report presents summary observations and insights gained from these four case studies.

2. East Bay Municipal Utility District

East Bay Municipal Utility District (EBMUD) is a public water utility established in 1923 under the California Municipal Utility District Act. Within the EBMUD service area, Special District Number 1 (SD1) was established in 1944 to treat wastewater.

2.1. Background

EBMUD provides water to an estimated 1.3 million people in 35 communities in Alameda and Contra Costa Counties in East San Francisco Bay, as well as industrial and commercial water users (Wallis et al., 2008; EBMUD, 2009b). It produces an average of 220 million gallons per day (mgd) of drinking water in non-drought years. The total service area is approximately 335 mi^2. EBMUD also provides wastewater services for approximately

640,000 customers west of Oakland/Berkeley Hills (EBMUD, 2009b) in an 83-mi^2 component of the EBMUD service area.

Diverse topography and maritime influences in California and the San Francisco Bay area contribute to a varied climate within the EBMUD service area (Figure 2). The Coast Range runs parallel to the coastline from Oregon to north of the Los Angeles Basin and is generally no more than 50 miles wide (WRCC, 2010). A break in the Coast Range at San Francisco Bay allows the inflow of marine air to the interior of the State under specific circulation patterns (WRCC, 2010). The Coast Range merges with the Cascade Range in the northern part of the State creating a 200-mile-wide area of rugged terrain (WRCC, 2010). The Cascades then tend southeast and merge into the Sierra Nevada, which continues to parallel the coast. Between these two ranges is the Central Valley. This flat, 45-mile-wide valley is closed off by the meeting of the Sierra Nevada and Tehachapi Mountains, which tend southwest to meet the Coast Range (WRCC, 2010).

West of these mountain ranges, the climate is predominantly maritime, dominated by the Pacific Ocean. This area experiences warm winters, cool summers, small daily and seasonal temperature ranges, and high relative humidity (WRCC, 2010). East of the mountain ranges, the climate is continental desert, characterized by warmer summers, colder winters, greater daily and seasonal temperature ranges, and generally lower relative humidity (WRCC, 2010). In the transition zone between these two areas, climate depends on how the local topography influences circulation patterns (WRCC, 2010). The difference between Oakland, California, on the San Francisco Bay, and Livermore, California, just 30 miles inland, illustrates the climate variability within the EBMUD service area. The average maximum July temperatures are 72°F and 89°F in Oakland and Livermore, respectively (WRCC, 2010).

Snowmelt from the Sierra Nevada feeds most major streams well into or throughout the arid summer months. Dams serve a dual purpose of providing a water supply throughout the dry part of the year and flood control during the winter and spring. In Oakland, the average total precipitation is 23 inches per year, while in Livermore it is 14 inches per year (NCDC, 2010). All of the precipitation in Oakland falls as rain while Livermore, on average, receives approximately 0.1 inch of snow (NCDC, 2010) annually.

Climate change has been documented in this region. In the second half of the twentieth century, a 3.6°F (2°C) rise in winter temperature was observed in the Sierra Nevada (EBMUD, 2009a). With a 9°F (5°C) rise in temperature, the

April 1 snow-covered area could decrease by as much as 50% (California Department of Water Resources [CA DWR] Report).

2.2. Description of the Water System

2.2.1. Drinking Water Supply System

2.2.1.1. Water Sources

The main water source for EBMUD is the Mokelumne River Watershed, which is located approximately 100 miles northeast of the service area in the Sierra Nevada (Figure 2). Approximately 90% of the water supply originates from this 577-mi^2 area (Wallis et al., 2008). The remaining water supply is from runoff in protected watershed areas of the East Bay into terminal reservoirs. During dry years, evaporation from East Bay terminal reservoirs can exceed runoff, resulting in no net water supply from local runoff in those years (EBMUD, 2009b).

Most of the Mokelumne River Watershed is undeveloped. Approximately 75% of the watershed is forested and located within national forests. Precipitation is highly variable in the watershed; 14 of the past 20 years have had below-normal precipitation or have been critically dry. Precipitation also varies considerably by season, with most precipitation occurring from November to May and the least from June to September. Peak flows take place during winter storms and the spring snowmelt; minimum flows occur in the late summer and fall (EBMUD, 2009b). Approximately 63% of the annual average runoff happens during the spring snowmelt from April to July (EBMUD, 2009a).

Two reservoirs on the Mokelumne River provide water storage, flood protection, recreation, hydropower, and resource management for a downstream fish hatchery. Flow into Pardee Reservoir is regulated by several upstream reservoirs. Pardee Reservoir has a maximum storage capacity of 197,950 acre-feet. The Mokelumne Aqueducts (three closed-pipe aqueducts) stretch 91 miles across the Sacramento/San Joaquin River Delta to convey water from the Pardee Reservoir to the EBMUD service area. The remaining water from the Pardee Reservoir flows to the Camanche Reservoir, which has a maximum storage capacity of 417,120 acre-feet. Water from the Pardee Reservoir is used to meet the demands of the EBMUD service area, while the Camanche Reservoir is managed to meet EBMUD's obligations to downstream fisheries and senior water rights (EBMUD, 2009b).

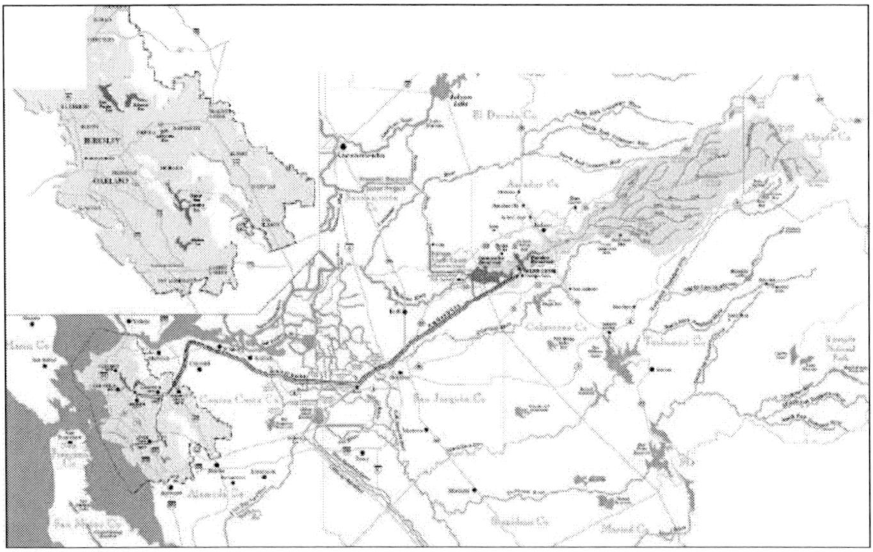

Source: EBMUD (2009b).

Figure 2. East Bay Municipal Utility District (EBMUD) service area and ultimate service boundary.

EBMUD has water rights and capacity to use or divert to storage up to 325 mgd of water from the Mokelumne River. The actual flow that can be diverted, however, is determined by the amount of runoff and streamflow, upstream and downstream senior water rights, and storage capacities. Additionally, the Camanche Reservoir must also provide releases for fisheries downstream and ensure the availability of up to 200,000 acre-feet of flood control storage during winter months (EBMUD, 2009b). Five terminal reservoirs have a combined capacity of 155,150 acre-feet (EBMUD, 2007). In addition to storing water from the Pardee Reservoir, the terminal reservoirs in the East Bay capture runoff from protected areas of the East Bay Watershed. The terminal reservoirs are operated to maintain 180 days of raw-water supply (EBMUD, 2009b).

Two additional water sources will be available starting mid-2010 to supplement water supplies during dry years (Chan, 2010). Up to 100 mgd of raw surface water will be available from the Sacramento River via the Freeport Regional Water Project. This additional water will meet approximately 22% of the need during dry years. EBMUD estimates that it will use this water source approximately 3 of every 10 years (EBMUD, 2009a). The other new source will be from the first phase of the Bayside Groundwater Project. Treated drinking water from the Mokelumne River will be injected into the south East Bay Plain Basin during wet years and extracted during dry years. The withdrawal permit provides for up to an annual maximum of 1 mgd of water with an extraction rate of 2 mgd for a portion of a "particular drought year" (EBMUD, 2009b).

2.2.1.2. Water Distribution

The water distribution system is composed of approximately 120 pressure zones (located at elevations ranging from sea level to 1,450 ft) and approximately 4,100 miles of pipe. About half of the water is distributed by gravity flow. In addition, there are approximately 140 pumping plants and 170 treated water storage tanks (EBMUD, 2007).

Water conveyed to EBMUD either is treated at one of three inline-filtration treatment plants and distributed or is stored in the East Bay terminal reservoirs. Three additional drinking water treatment plants are supplied by two terminal reservoirs. These three plants have full conventional treatment, with two of them also providing ozonation.

2.2.1.3. Water Use

Water use in the EBMUD service area is approximately 92% residential, 7% commercial, and 1% industrial and public authority use (EBMUD, 2007). Most water provision services are funded by user fees (approximately 75%) with the remaining revenue coming from capital contributions, investments, taxes, hydropower generation, and other sources (EBMUD, 2009c).

2.2.1.4. Demand Management

Programs for managing demand include water rationing, conservation, and reuse. In calculating water availability, EBMUD follows its Water Supply Availability and Deficiency Policy. According to this policy, the maximum rationing (i.e., mandatory water use reduction) during droughts is a 25% reduction in total customer demand, while continuing to provide water to fisheries and other downstream obligations (EBMUD, 2009b). Varying levels of rationing are imposed, depending on the existing and projected extent of the drought and how the levels differ across customer categories. Conservation measures include leak detection and repair in the distribution system, customer incentives for water reduction, and customer education and outreach on water conservation. EBMUD reuses water by providing treated wastewater and untreated raw water from local runoff for irrigation and in-plant processes (EBMUD, 2009b). Approximately 9.3 mgd of water is recycled (Towey, 2010).

2.2.2. Wastewater System

EBMUD provides wastewater treatment in SD1, a subset of the water service area. Nine communities within SD1 have wastewater collection systems that discharge into one of EBMUD's five interceptor sewer trunk lines (EBMUD, 2010). The interceptors have a capacity of 760 mgd of water. On average, the EBMUD wastewater treatment plant (WWTP) in Oakland receives 80 mgd from the interceptors (EBMUD, 2007). The Oakland WWTP has the capacity for 320 mgd of primary treatment, 168 mgd of secondary treatment, a short-term hydraulic peak of 415 mgd during wet weather events, and 11 million gallons of storage (EBMUD, 2007; Cheng, 2010). Treated wastewater is discharged 1 mile off the coast through a deep-water outfall into San Francisco Bay (LAFCO, 2008; EBMUD, 2007).

By-products from WWTP operations are used in two forms: Biosolids are used as a soil amendment or alternative daily cover at landfills, and methane gas provides energy needed for operations (EBMUD, 2007). Additionally, as part of its wastewater source control and pollution prevention activities, EBMUD collects concentrated domestic waste, oil, and grease from restaurants, and other highly organic waste streams to produce methane gas, while decreasing the organic content of the wastewater stream (EBMUD, 2007). Overall, self-produced methane gas provides up to 90% of the Oakland WWTP's power supply (Cheng, 2010).

Since 1979, EBMUD and local communities have addressed rainwater infiltration and inflow in the wastewater collections system resulting from deteriorated pipes and improper storm drain connections. As part of the East Bay Infiltration/Inflow Correction Program, EBMUD constructed three wet-weather treatment plants, two storage basins, 7.5 miles of new interceptor lines, and an expanded Oakland WWTP. Communities have spent more than $460 million on improvements for their wastewater collection systems (EBMUD, 2007).

In 2009, approximately 69% of the revenue for wastewater services came from user fees (53% from wastewater, 16% from wet-weather facilities), and the remaining came from capital contributions, resource recovery, taxes, investments, and other sources (EBMUD, 2009c).

2.3. Climate Change Projections and Risks

The climate change information EBMUD used to evaluate vulnerability to climate change included the *Fourth Assessment Report of th Intergovernmental Panel on Climate Change* (IPCC, 2007) and two state-level studies that modeled the effects of climate change on water resources (EBMUD, 2009a). Model projections from the IPCC suggest that temperatures in the western United States could rise 3.6−13.5°F (2.0−7.5°C) by the end of this century (IPCC, 2007, as cited in Wallis et al., 2008). In a summary of northern California climate change studies, Dettinger (2004, as cited in EBMUD, 2009a) provides a range of a 3.6−10.8°F (2.0−6.0°C) increase in temperature and either a 20% increase or decrease in precipitation. Rising temperatures are expected to cause precipitation to fall more often as rain, decreasing water storage in snowpack and causing spring runoff to occur earlier. The temperature rise will extend the growing season by about 19−28 days, with more frequent and longer heat waves (Wallis et al., 2008). Sea level is expected to rise another 0.6−1.9 ft by the end of the century (IPCC, 2007, as cited in Wallis et al., 2008). This rise in sea level will affect the frequency and severity of flooding in coastal areas, including the flood-prone Sacramento/San Joaquin River Delta, which three EBMUD water transmission aqueducts cross (Wallis et al., 2008).

EBMUD reviewed two state-level climate change studies—one by the California Energy Commission's Public Interest Energy Research (PIER) and the California Climate Change Center (CCCC), and one by the California Department of Water Resources (CA DWR). A review of both state-level studies by EBMUD concluded that the studies yielded the following similar but uncertain results (EBMUD, 2009a):

- Temperature increases will be significant, but the magnitude of change is uncertain.
- Snowpack volume will decrease.
- Snow will melt earlier.
- The direction and amount of change in total annual precipitation are inconclusive.
- Drought impacts are inconclusive, but some scenarios predict increased frequency and longer duration droughts.
- Climate variability will generally increase.

With a growing awareness of climate change and its potential effects on water resource management, EBMUD started following climate change research, collecting information about projected regional climate change, gathering environmental data, and networking locally and nationally with others in the water community (Wallis et al., 2008; Chan, 2010). EBMUD staff presented these efforts to the Board of Directors and at an annual business forum the Board of Directors and key stakeholders attended (Wallis et al., 2008).

Additionally, EBMUD gauged customer opinion about climate change in an annual customer survey. Of the respondents, almost 75% thought that climate change will be an issue for water suppliers within the next 50 years, and the effect of climate change on water availability was of "highest concern" for 46% (Wallis et al., 2008; Chan, 2010).

In mid-2007, EBMUD established an official utility-wide management approach for addressing climate change and formed a cross-departmental climate change committee. The committee's primary tasks include keeping up to date on climate change science, evaluating the potential effects of climate change, reviewing Mokelumne River Watershed data to determine changes in trends, assessing water supply and infrastructure vulnerabilities, integrating climate change in planning and budgeting, and developing adaptation and mitigation strategies. By 2008, the EBMUD strategic plan added climate change as one of the strategies for meeting long-term water supply goals (EBMUD, 2008). Strategies included developing and implementing a Climate Change Monitoring and Response Plan and mitigating greenhouse gas emissions across departments (Wallis et al., 2008). While climate change-related activities, such as mitigating greenhouse gas emissions are cross-departmental, vulnerability assessment efforts have focused primarily on the water supply system (Cheng, 2010).

2.4. Climate Change Vulnerability Assessments

EBMUD identified four key areas of potential vulnerability to climate change: (1) flooding and sea level rise, (2) hydropower generation, (3) water supply and demand, and (4) water quality (Wallis et al., 2008). Since 2006, EBMUD has conducted qualitative assessments and sensitivity analyses to examine these vulnerabilities and their impacts on the drinking water system.

The most extensive and quantitative vulnerability analysis was completed as part of the Water Supply Management Program (WSMP) 2040. Vulnerability analyses for the WSMP 2040 focused on water supply, water demand, and the effect of temperature on water quality. Qualitative and less formal assessments have been performed for flooding, sea level rise, and power generation. EMBUD also participates in local and national conferences and workgroups, such as the U.S. Environmental Protection Agency (EPA) Climate Ready Water Utilities Working Group, and currently is working with the U.S. EPA and the Water Research Foundation on developing vulnerability and risk assessment tools to assist other water utilities in conducting climate change analyses (Chan, 2010). For more detailed information on how EBMUD and other utilities organized their vulnerability assessments, see *Climate Change Vulnerability Assessments: A Review of Water Utility Practices* (U.S. EPA, 2010).

2.4.1. Flooding and Sea Level Rise

EBMUD expects that flooding could increase as a result of the more frequent extreme weather events that are predicted with climate change. To assess the effect of more extreme weather events on the potential for flooding in urbanized areas downstream of the Camanche Reservoir, EBMUD modeled the water supply system with a 5.4°F (3°C) rise and 1997 precipitation levels (the wettest year in the past quarter century due to El Niño). The study used the daily operational model for the EBMUD water system (Chan, 2010). Results showed that the peak water release from the Camanche Reservoir would have had to be three times greater than it was in 1997 to prevent riverine flooding (Wallis et al., 2008).

In addition to more extreme weather events, sea level rise could contribute to increased coastal flooding. A 1-foot rise in sea level could cause the 1-in-100-year storm surge flood event to occur once every 10 years (Wallis et al., 2008). The aging levee system in the flood- and earthquake-prone Sacramento/San Joaquin River Delta is an existing vulnerability that would be exacerbated by rising sea levels. Such coastal flooding could disrupt water delivery for months as it did in 2004, when a single levee breach caused flooding that submerged the aqueducts for more than 4 months.

As part of WSMP 2040, EBMUD reviewed the two state-level climate change studies (Section 2.3, above) and found that they sufficiently document current conditions and existing risks, including the susceptibility of the raw-water system to levee failures, earthquakes, and potential failure scenarios.

The interactions of vulnerabilities, such as the effects of sea level rise on levee failure, however, have not been characterized. CA DWR is drafting a Delta Risk Management Strategy, and its first report will provide discrete probabilities of levee failure considering several climate change and sea level-rise scenarios. EBMUD plans to use this information to comment on improvement options CA DWR proposes (EBMUD, 2009b).

2.4.2. Hydropower Generation

Although extreme weather events could cause more intense precipitation and flooding, total annual precipitation might decrease. Decreased annual precipitation would not only affect the ability to meet water needs but also would affect hydropower generation. To model the potential range of effects, EBMUDSIM was used. EBMUDSIM is a monthly time-step model of the EBMUD water supply system from the Mokelumne River reservoirs to the five terminal reservoirs in the service area, all of which are modeled as one combined reservoir (Chan, 2010). Results suggested that the projected changes in total precipitation could lead to a 10–30% decrease in hydropower generation (Wallis et al., 2008).

2.4.3. Water Supply

EBMUD has had several ongoing activities related to climate change, but the first extensive, quantitative analyses to assess the effects of climate change on its water supply system were conducted for WSMP 2040. The main objective of WSMP 2040 was to identify and recommend a portfolio of projects for meeting customers' dry-year water needs through 2040.[4] The process consisted of six steps: (1) identifying a list of projects that could provide additional supply, (2) screening the projects, (3) developing portfolios of projects that satisfy water needs through 2040, (4) screening 14 preliminary portfolios under historic hydrologic conditions with existing drought planning sequence, (5) modeling five of these portfolios under the effects of projected climate change, and (6) making a final portfolio selection. Projects included changes in rationing, conservation, water reuse, surface water transfers, groundwater banking/exchange, desalination, and enlargement of reservoir(s) (EBMUD, 2009b). Several uncertainties were identified regarding the proposed projects, including institutional and legal challenges, undefined timelines for project completion, and climate change. To reduce these uncertainties, a reliable portfolio was defined as being (1) robust with respect to an uncertain future, (2) composed of projects that can be pursued simultaneously, and (3) flexible and diverse (EBMUD, 2009c). To inform the selection of a reliable portfolio, a climate change vulnerability analysis was conducted.

EBMUD reviewed 10 other water agencies in California to determine how each was assessing its vulnerabilities to climate change (EBMUD, 2009a). Based on this information, EBMUD considered five approaches for evaluating the effects of climate change on the water supply system: (1) qualitative analysis, (2) perturbing historic hydrology based on perturbation factors from existing studies, (3) hydrologic modeling based on existing climate-derived hydrology by other studies, (4) hydrologic modeling using climate-derived temperature and historic precipitation, and (5) sensitivity analyses using historic hydrology in a hydrologic model (EBMUD, 2009b). A "bottom-up" approach using sensitivity analyses was selected based on a recommendation by Miller and Yates (2006). A bottom-up approach consists of identifying the factors that most affect the system's reliability and testing that sensitivity to and performance under expected changes in those factors (EBMUD, 2009b).

EBMUD identified the three most significant factors that affect the water supply system's reliability in meeting the projected 2040 dry-year water needs: (1) greater-than-expected customer demand, (2) shift in the timing of spring runoff, and (3) decreased volume of precipitation and runoff. EBMUD modeled three sets of scenarios based on these three factors with potential changes in each factor based on the existing regional climate change studies to determine the effect of each factor on the performance of the existing system.[5] Modeling assumptions included using existing conservation and recycled water levels, existing drought planning sequence, and a maximum of 25% rationing. The model was run from 1953 to 2002 according to each of the three scenarios (EBMUD, 2009a). Although climate change projections from the IPCC, PIER/CCCC, and CA DWR reports have significant uncertainties, they provided an approximate range of potential changes in climate and hydrology. From this range, EBMUD selected and modeled those changes that are expected to affect the utility's ability to provide sufficient water and meet regulatory obligations for downstream water flow and temperature (e.g., increases in precipitation were not modeled).

2.4.3.1. Temporal Shift in Runoff

As a result of increasing temperatures, the volume of runoff between April and July has decreased by approximately 10% over the past century (Wallis et al., 2008). The sensitivity of the water supply system to reductions in spring runoff and increased winter runoff was modeled for 3.6°F, 5.4°F, and 7.2°F (2°C, 3°C, and 4°C) increases in temperature. The analysis estimated the decrease in the volume of runoff from April through June and assumed an increase in the November-to-March runoff by the same volume (EBMUD, 2009a). With 3.6°F, 5.4°F, and 7.2°F (2°C, 3°C, and 4°C) temperature increases, estimated reductions in April through June runoff were 19%, 28%, and 38%, respectively. Carryover storage decreased by an average of 2.5–6% and a maximum of 10–16%. Customer rationing was estimated to increase by a maximum of 7%. Flood control releases increased in 60% of the years between November and May by an average of 66–89%. Between April and July, flood control releases decreased in 35% of the years by 40–80% (EBMUD, 2009a).

The shifts in the timing of runoff had no significant impact on EBMUD's ability to meet water demand because EBMUD's total reservoir storage is larger than the total annual average runoff (EBMUD, 2009a). This capacity enables system operations to be reconfigured for fewer flood control releases (Wallis et al., 2008). When considered in combination with the need to adjust flood releases from the Camanche Reservoir to accommodate extreme precipitation events as predicted above, however, the finding suggests a more delicate balance between flood control and capturing the projected temporal shift in spring runoff.

2.4.3.2. Decrease in Annual Precipitation

The effect of reduced precipitation was assessed by assuming that reductions of 10% and 20% in the volumes of annual precipitation directly correspond to 10% and 20% decreases in runoff. Both scenarios were run in the W-E model, a linked combination of the Water Evaluation and Planning and EBMUDSIM models, which resulted in the most significant effects observed among all the scenarios. For the 10% and 20% reductions in precipitation, the average decreases in carryover storage were 12% and 24%, respectively, and the maximum decreases were 47% and 76%, respectively. The magnitude of customer rationing increased on average by 3.8% and 6.4% for the 10% and 20% decreases in precipitation, respectively. The frequency of rationing increased from a baseline of 36% to 44% and 52%, respectively, for the 10% and 20% decreases in precipitation. Average annual flood releases decreased by 43% and 74% for the 10% and 20% decreases in precipitation, respectively (EBMUD, 2009a).

In comparison, the worst drought on record occurred in 1976 and 1977 and resulted in a 75% decrease in average runoff and a 70% reduction in total reservoir capacity (EBMUD, 2009a). A limitation of these sensitivity analyses is that the change in each vulnerability factor was modeled individually, and the synergistic effect of a simultaneous change in all three factors at same time was not considered (EBMUD, 2009b). A final scenario with all three factors would have provided insight into the worst-case scenario.

2.4.3.3. Increased Demand

To test the effects of increased water demand, 2040 water demand estimates were recalculated assuming a 7.2°F (4°C) increase in air temperature, resulting in a 3.6% increase in demand.[6] The higher demand estimate accounts for higher consumptive use for drinking and outdoor watering due to higher temperatures alone. A 20% decrease in precipitation had relatively little effect on demands compared to the temperature increase. Therefore, only the demand estimate based on a temperature increase was run in the W-E model. Results showed an average decrease in carryover storage of 3%, with a maximum decrease of 8%. Carryover storage is significant for the EBMUD water supply system, because the reservoirs do not necessarily refill each year, depending on drought conditions. The results also indicate that the magnitude of customer rationing increased to a maximum of 5.6% of demand, but the frequency with which rationing occurred did not change. Flood control releases were not evaluated in this analysis (EBMUD, 2009a).

2.4.4. Water Quality

EBMUD used the Watershed Analysis Risk Management Framework (WARMF) model to assess the effect of increasing air temperatures on water temperatures. The WARMF model was developed by the Upper Mokelumne River Watershed Authority for a different study in which EBMUD participated. This previous analysis was completed to determine the effect of climate change on EBMUD's continued ability to meet its cold-water obligations to the downstream fish hatchery (EBMUD, 2009a).

Six water years were modeled, including two dry years, three below-normal years, and one above-normal year. Each year was modeled for increases of 3.6°F, 5.4°F, and 7.2°F (2°C, 3°C, and 4°C) rise in air temperature projected to occur from 1980 to 2040. Over all scenarios, average annual water temperatures increased by 0.5−6.3°F (0.3−3.5°C) relative to baseline temperatures. In general, the effect of increasing temperatures was found to depend on the type of hydrologic year and the season. In drier years and during the summer months, streamflow is lower and air temperatures have a greater effect on water temperatures (EBMUD, 2009a).

EBMUD studies also identified other water quality effects from climate change, including a greater potential for algal growth with higher water temperatures. In addition, with increasing intensity and frequency of storm events, turbidity levels can increase in water supply sources. Because the EBMUD drinking water treatment plants were designed for treating source water that is low in turbidity, increases in turbidity could decrease the plants' daily outputs and increase treatment costs (Wallis et al., 2008).

2.5. Application of Vulnerability Assessment Information

Several key insights were provided by the climate change analyses described above for the WSMP 2040 decision-making process. The analyses showed a clear distinction between the effects of temporal shifts in precipitation and a decrease in total annual precipitation. The temporal shifts could be managed by adjusting system operations, while decreased precipitation would require additional sources of water outside of the Mokelumne River Watershed. Before conducting these studies, EBMUD believed that diversification of water supply sources was needed. The studies reaffirmed the need for diversifying water supply sources outside of the watershed and selecting projects that can be adapted as climate change effects are observed. For example, instead of relying solely on enlarging existing reservoirs, EBMUD will pursue additional surface water and groundwater sources. Plans also will be drawn up for regional desalination. To meet the 2040 dry-year water needs, conservation, desalination, and the enlargement of reservoirs in combination with groundwater banking and exchange are needed. Pursuing parallel tracks on alternative projects will allow for flexibility, not only with regulatory and logistical challenges, but also with adjusting to future refinements of climate change projections. Although water quality vulnerabilities were not directly addressed, the analyses revealed that the interaction of lower water levels in the reservoirs and increased air temperatures are causal factors for potential water quality concerns.

To support continued climate change vulnerability assessments and adaptation activities, EBMUD identified two main resources that would support these efforts: (1) information on the probabilities of specific projected changes in temperature and precipitation, and (2) a common source for region-specific environmental data to assist in vulnerability analyses (Chan, 2010).

3. New York City Department of Environmental Protection

3.1. Background

New York City's Department of Environmental Protection (DEP) is responsible for the operation, protection, and maintenance of New York City's drinking water system (DEP, 2010a). DEP supplies 1.1 billion gallons per day (gpd) of drinking water to 8.2 million residents of New York City, and an additional 1 million people in nearby municipalities (DEP, 2008a; see Figure 3). DEP supplies approximately 85% of the water for Westchester County and 5−10% of the water for Orange, Putnam, and Ulster Counties (Rosenzweig et al., 2007). The system also provides legally mandated conservation releases to the Delaware River Basin (DEP, 2008a).

New York State experiences a humid continental climate but with dramatic variations from that climate type due to latitude, general circulation patterns, and topography. Although the region is located along the coast, the area is dominated by drier continental airflow from the prevailing westerly winds. The state's climate is conditioned primarily by cold, dry air masses from the northern continental interior, as well as warm, humid air masses from the south conditioned by the Gulf of Mexico. A third, but relatively less important influence is the air mass from the North Atlantic Ocean, which can produce cool, cloudy, and damp weather. Due to the prevailing winds, however, this maritime influence is secondary to the more prevalent airflow from the continental interior (New York State Climate Office, 2010).

Average annual temperature is approximately 55°F in New York City but 10−15°F cooler in the Catskills, which is the major source of water. The distribution of precipitation across New York State is influenced by topography and proximity to the Great Lakes and Atlantic Ocean. Average annual precipitation can exceed 50 inches in the Catskills. In New York City, average annual precipitation is 43−50 inches per year, depending on location within the city. Precipitation is relatively evenly distributed throughout the year, and no distinctly wet or dry seasons occur (NYC Panel on Climate Change, 2009; New York State Climate Office, 2010).

In the mountainous areas of New York State, such as the Catskills, annual average snowfalls range from 70−90 inches, but topography and elevation cause snowfall over even short distances in the state's interior to vary greatly. The bulk of wintertime precipitation in these areas falls as snow. New York

City, however, receives only about 25–35 inches of snow per year due to the moderating influence of the Atlantic Ocean. Because of the temperature modulation of the coastal zone, only about one-third of the winter-season precipitation is associated with storms having snow accumulation of at least 1 inch (New York State Climate Office, 2010); the rest falls as rain.

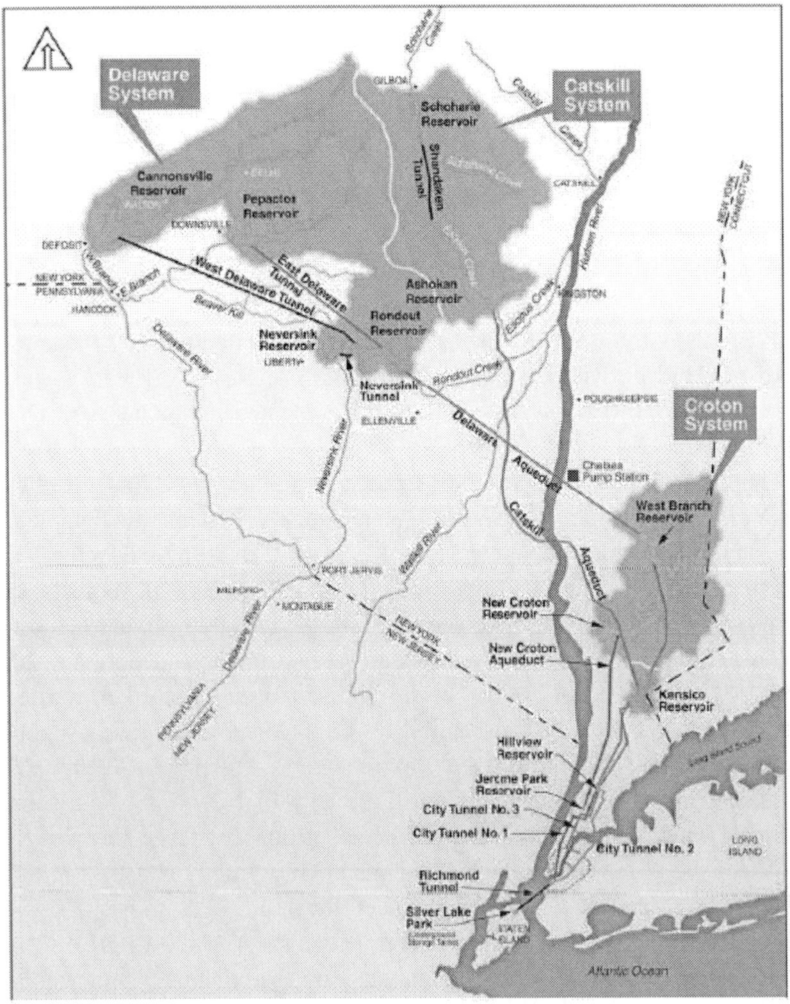

Source: DEP (2008b).

Figure 3. New York City Department of Environmental Protection (DEP) system overview.

Annual mean temperature in New York City has increased 2.5°F since 1900, although both warming and cooling periods occurred over this time. Mean annual precipitation levels have increased only slightly since 1900, but year-to-year variability in precipitation has increased (NYC Panel on Climate Change, 2009).

3.2. Description of the Water System

3.2.1. Water Supply System

New York City's surface water is supplied from a network of 19 reservoirs and three controlled lakes in a region that stretches nearly 2,000 mi^2 and extends 125 mi north of the City (Figure 3). This region is divided into two geographically discrete systems—the Catskill/Delaware reservoir systems, located in upstate New York, well north and west of the City and the Hudson River, and the Croton Reservoir system, which is located north of the City and east of the Hudson River. Some (less than 1%) of New York City's water is obtained from the Brooklyn-Queens Aquifer, located in southeastern Queens (DEP, 2008b).

The Catskill/Delaware reservoir systems provide approximately 90% of New York City's water. The Catskill Water Supply System was completed in 1927, and the Delaware Water Supply System in 1967. Together, these watersheds cover roughly 1,600 mi^2 (U.S. EPA, 2009b). Forests cover approximately 75% of the watersheds. More than 20,000 private landowners own an estimated 75% of the forested land area (Brunette and Germain, 2003).

In 1993, New York City began implementing watershed protection programs to reduce the susceptibility of the surface water supply to contaminants. In 1997, The U.S. Environmental Protection Agency (EPA) partnered with the State of New York, the City of New York, and some 80 watershed municipalities and environmental groups to forge the New York City Watershed Memorandum of Agreement (MOA). This MOA set forth a set of conditions that the City had to meet for U.S. EPA to issue a 5-year Filtration Avoidance Determination (FAD). This FAD allows DEP to avoid filtering its Catskill/Delaware drinking water by establishing a land acquisition program for source water protection, by setting more stringent New York City watershed rules and regulations, and by implementing other watershed protection strategies. U.S. EPA reissued New York City a 5-year FAD in 2002 and a 10-year FAD in 2007. The New York State Department of Health and the U.S. EPA monitor these ongoing source water quality programs. Projects include

- *Land Acquisition*—New York City buys property from willing sellers to buffer the reservoirs and controlled lakes.
- *Land Management*—DEP develops land management programs.
- *Partnership Programs*—DEP partners with many local organizations to help protect source water quality, for example, by improving septic systems.
- *Wastewater Treatment Plant Upgrades*—New York City funds improvements to non-city-owned wastewater treatment plants for communities in the source watersheds.
- *Stream Management Programs*—DEP supports partnerships to stabilize streams in the area.
- *Watershed Agricultural Programs and Forestry Program*—These programs work with farms to help implement best management practices that reduce agricultural pollution and protect water quality (DEP, 2008b).

Glacial clay deposits underlay stream channels and steep topography surrounds the waterways in the Catskill water system. As a result, the system is prone to high turbidity due to intense precipitation events and associated runoff. Maintaining the FAD for the Catskill and Delaware water supplies is a crucial element to future watershed plans. To meet FAD-required standards, DEP has occasionally added alum to the waters entering Kensico Reservoir to reduce turbidity.[7] Periodically, the alum and associated sediment must be dredged from the reservoirs (DEP, 2005).

The Croton Watershed contains three upland reservoirs which supply approximately 10% of the City's water. It covers roughly 375 mi^2 east of the Hudson River in Westchester, Putnam, and Dutchess Counties and a small section of Connecticut. The Croton system began service in the mid-1800s and was completed before World War I (Rosenzweig et al., 2007). Since the 1950s, the Croton Watershed has developed quickly with the construction of 60 wastewater treatment plants, interstate highways, residential developments, and impervious surfaces (New York Water, 2010). On several occasions, the Croton Watershed has been contaminated as a result of stormwater runoff.

For example, 12 of Westchester county's 45 municipalities lie within the boundaries of the Croton Watershed. These municipalities contribute to water supply contamination as a result of lawn-care chemical use, automobile use, combined sewer system overflows, and other human factors, as well as reduced infiltration of precipitation that flows through urban drainage infrastructure. In 1993, the U.S. EPA determined that the Croton system failed to meet the requirements of the Surface Water Treatment Rule, and Croton system raw water would need to be filtered and disinfected. Repeated violations of turbidity and disinfectant by-product rules under the 1996 Safe Drinking Water Act Amendments have caused DEP to periodically remove the Croton system from service (Water-Technology.net, 2010; DEP, 2010b). After several delays and consent orders resulting in fines, the first phase of construction of the Croton raw-water treatment plant began in 2006 and is expected to be operational by 2012. Treatment will include a pretreatment stage, mixing and coagulation, flocculation, chemical balancing, stacked dissolved air floatation, and ultraviolet and chlorine treatment. The filtration plant is expected to improve water quality by reducing turbidity, decreasing the risk of microbiological contamination, and reducing the levels of disinfection by-products (DEP, 2008b). Communities around the Croton Watershed also signed the 1997 MOA aimed at improving watershed protection. They are participating in land acquisition and other raw water quality projects, as discussed above (DEP, 2008b).

3.2.2. Wastewater System

DEP is also responsible for operating, protecting, and maintaining New York City's wastewater system. The wastewater network includes more than 6,000 mi of wastewater pipes, 135,000 sewer catch basins, 495 permitted outfalls, 95 wastewater pumping stations, and 14 wastewater treatment plants spread across the City's 5 boroughs (DEP Web site). On average, the system treats 1.4 billion gpd of wastewater and has the capacity to treat dry-weather flows of 1.8 billion gpd (DEP, 2006).

New York City's wastewater undergoes five major processes: preliminary treatment, primary treatment, secondary treatment, disinfection, and sludge treatment. Approximately 60% of the City's sewers are combined, making combined sewer overflows (CSOs) a continuing problem for DEP (DEP, 2008a) during intense precipitation events. Violations of New York City's 1988 State Pollutant Discharge Elimination System permit led to a 1992 consent order between New York State's Department of Environmental Conservation and DEP, requiring a CSO abatement program. A 2004 consent order with more detailed guidance includes requirements for more than 30 citywide projects, such as sewer separation, flushing tunnels, off-line retention tanks, and vortex concentrators to improve the efficiency of the wastewater system (NYSDEC, 2010). While continuing to invest in traditional, or grey, infrastructure, the City is implementing measures to maximize the use of green infrastructure and other source controls to reduce stormwater runoff from new and existing development. Unveiled in September 2010 and subject to regulatory negotiations and approvals, the NYC Green Infrastructure Plan marks a departure from conventional and expensive approaches to stormwater management.

3.3. Climate Change Projections and Risks

DEP expects temperatures in New York City and its watersheds to increase by 1.5–3°F by the 2020s, 3–5°F by the 2050s, and 4–7.5°F by the 2080s (Table 2). The variability in precipitation in the New York City area is large. Most climate model projections indicate small increases in precipitation, but some suggest precipitation decreases, thus reducing confidence in projections of precipitation in this region. The New York City Panel on Climate Change concluded in 2009 that the best estimates at this time indicate approximate increases of 0–5% by the 2020s, 0–10% by the 2050s, and 5–10% by the 2080s. Most models indicate precipitation increases for the winter months and slight decreases during September and October. Furthermore, as temperatures increase, more precipitation is expected to fall as rain instead of snow (NYC Panel on Climate Change, 2009). In short, the observed climate change trends are projected to continue and, in some cases, accelerate.

New York City has taken a proactive approach to climate change. In 2001, the City joined the Local Governments for Sustainability's Cities for Climate Protection campaign. In 2004, DEP created a climate change task force to assess the potential impacts of climate change on the City's water infrastructure. The task force comprises representatives from a variety of DEP's offices and initially included participants from Columbia University's Center for Climate Systems Research, the National Aeronautics and Space Administration's Goddard Institute for Space Studies, HydroQual Environmental Engineers and Scientists, P.C., the New York City Office of Environmental Coordination, the Mayor's Office of Long-term Planning and Sustainability, and the New York City Law Department. The task force created an action plan, which includes the following tasks (DEP, 2008a):

- Work with climate scientists to improve regional climate change projections
- Enhance DEP's understanding of the potential impacts of climate change on DEP's operations
- Determine and implement appropriate adaptations to DEP's water systems
- Inventory and manage greenhouse gas emissions
- Improve communications and tracking mechanisms.

A sustainability plan for New York City, PlaNYC, was unveiled on Earth Day in 2007. The plan outlines a 25-year vision for the city, focusing on maintaining and improving the City's infrastructure centering on land, water, transportation, energy, air, and climate change. PlaNYC has set an ambitious target to reduce the City's greenhouse gas emissions by 30%. New York City's plan for climate change adaptation includes (1) creating an intergovernmental task force to protect the City's infrastructure, (2) working with vulnerable neighborhoods to develop site-specific plans, and (3) launching a citywide strategic planning process (PlaNYC, 2007a; PlaNYC, 2007b; PlaNYC, 2008-2010).

To respond to climate change in New York City and to meet the goals established in PlaNYC, the New York City Panel on Climate Change (NPCC) was created in 2008. This panel is composed of climate change scientists and legal, insurance, and risk management experts. With funding from the Rockefeller Foundation, NPCC has been charged with serving as the technical advisory body for the Mayor and the New York City Climate Change Adaptation Task Force. This organization has provided the Climate Change Adaptation Task Force with the most comprehensive set of climate data that has been produced for New York City (NYC Panel on Climate Change, 2009). Several experts engaged to assist DEP in 2004 also were enlisted to assist NPCC in citywide planning efforts. DEP continues to pursue complementary climate change research because it is concerned with climate change in upstate New York (where the Catskill and Delaware reservoir systems are located) and in the City itself.

3.4. Climate Change Vulnerability Assessments

DEP's vulnerability work is based on three core questions of interest to DEP, including the potential effects of climate change on (1) total water supply, (2) turbidity, and (3) eutrophication (Barsugli et al., 2009). DEP worked with researchers from Columbia University's Center for Climate Systems Research to design its Climate Impact Assessment project (Major et al., 2007). The goal of this integrated modeling project is to estimate the effect of future climate change on the quantity and quality of New York City's water supply. The project will use climate change projections, DEP water quality and water supply models, and analytical measures of system performance to advance DEP's understanding of the potential impacts of climate change on the water supply system. For more detailed information on how DEP and other utilities organized their vulnerability assessments, see the U.S. EPA report, *Climate Change Vulnerability Assessments: A Review of Water Utility Practices* (U.S. EPA, 2010).

The project was planned in two phases. Phase I, now completed under contract with Columbia University and the City University of New York (CUNY), is aimed to provide a first-cut evaluation of the effects of climate change on water quantity and quality in selected portions of the water system, using the existing modeling system and data available from three general circulation models (GCMs). Phase II, now in process with continued support from CUNY, has goals similar to those of Phase I but with upgrades to both models and data sets applied to the entire water supply system, including a greater variety of GCM data and an evaluation and application of differing downscaling methods. The Phase I effort used the Intergovernmental Panel on Climate Change (IPCC) *Third Assessment Report* (DEP, 2008a; McCarthy et al., 2001), but current efforts have upgraded models and data that were used in the IPCC *Fourth Assessment Report* (Parry et al., 2007).

A climate change scenario framework was developed for the New York City water supply system using high temporal resolution data from the Program for Climate Model Diagnosis and Intercomparison (PCMDI) Web site maintained by the Lawrence Livermore National Laboratory in Berkeley, California (Maurer et al., 2007). Data for Phase I were extracted from the single grid box at the center of the watershed region. Baseline data for 1981–2000 came from "hindcast" model runs. Data for 2046–2065 and 2081–2100 came from three GCMs (the Goddard Institute for Space Studies [GISS] Model, the Max Planck Institute [MPI] ECHAM5, and the National Center for Atmospheric Research [NCAR] CCSM3) coupled with three scenarios from the IPCC *Special Report on Emissions Scenarios*: *A1B, A2, and B1*. The data included mean temperature, maximum temperature, minimum temperature, precipitation, sea level pressure, zonal wind, meridional wind, solar radiation, longwave radiation, and dewpoint temperature.

Table 2. Projected baseline climate and mean annual changes for New York City

Climate indicators	Baseline 1971–2000	2020s	2050s	2080s
Air temperature (°F)	55	+1.5–3	+3–5	+4–7.5
Precipitation	46.5 in	+0–5%	+0–10%	+5–10%
Sea level rise (inches)	N/A	+2–5	+7–12	+12–23
Number of days/year with max temp. above 90°F	14	23–29	29–45	37–64
Number of days/year with max temp. above 100°F	0.4	0.6–1	1–4	2–9
Number of heat waves/year	2	3–4	4–6	5–8
Number of days/year with rainfall exceeding 1 inches	13	13–14	13–15	14–16
Number of days/year with rainfall exceeding 2 inches	3	3–4	3–4	4–4
Number of days/year with rainfall exceeding 4 inches	0.3	0.2–0.4	0.3–0.4	0.3–0.5

Source: NYC Panel on Climate Change (2009, pp.17, 20).

For Phase I, each scenario was used to calculate delta change coefficients representing mean monthly change in air temperature and precipitation between control and future prediction periods. Monthly delta change factors were applied additively for air temperature and as a ratio for precipitation to historical meteorological period data, generating a future prediction time series. The feasibility of applying the delta change method to the wind and solar radiation data needed for the reservoir models was also investigated.

For Phase II, GCM selection included the entire CMIP3 multi-model data set, the A2, A1B, and B1 emissions scenarios, and seven meteorological variables (precipitation; maximum, minimum, and average temperatures; zonal and meridional winds; and solar radiation). Data from all the GCMs were re-gridded to 2.5° corresponding to the Eastern North America region using bilinear interpolation and the NCAR Command Language (NCL: www.ncl.ucar.edu). The various levels of data processing involved necessitated that some data be eliminated from the study depending on the number of models that contain a given meteorological parameter, the number of runs archived for each GCM, and whether data existed for all points necessary in the re-gridding process. GCM hindcasts were compared to historical data sets at four spatial scales: Eastern North America, the nine grid points surrounding West of Hudson watersheds, the four grid points surrounding New York City, and the single grid point closest to the centroid of New York City watersheds. To develop a skill ranking and probability distribution function for each meteorological variable, spatial scale, season (December to February, March to May, June to August, and September to November), and GCM, the fidelity of hindcast values to observed historical data was calculated.

The system of models that will be used for the integrated modeling project include the General Watershed Loading Function (GWLF) and Soil Water Assessment Tool (SWAT) watershed models, a one-dimensional reservoir eutrophication model, a two-dimensional reservoir turbidity transport model (CEQUAL-W2), and the OASIS system operations model for the entire water supply. These models taken together with the existing and in-process climate scenarios make the proposed integrated assessment possible.

As the project progresses, further model enhancements and integration will be implemented. For the GWLF watershed model, improvements will be made to the following model elements: hydrologic balance, sediment and nutrient generation and transport, ecosystem effects, and land use. For the reservoir models, additional upgrades and calibration and development of response function models keyed on system performance measures will be implemented. For the integrated system, enhanced coupling of the watershed and reservoir models to OASIS will be undertaken. And for model inputs, enhancements will include advanced delta change with historical data morphing, statistical downscaling, and regional climate model (RCM) simulations.

Several performance measures related to water system quantity and quality will be developed and used to estimate climate change effects, including total water quantity, probabilities of refill, probabilities of drawdown, key point turbidity levels, frequency of alum use, reservoir phosphorus and chlorophyll concentrations, and restrictions in water use due to eutrophication. DEP expects the results of this project to provide the basis for recommendations about system operation now and in the future, and, in later phases, recommendations about required infrastructure changes and improvements.

In 2009, the NPCC published its first report, *Climate Risk Information.* This report provides climate change projections for New York City as a whole (not just DEP) and identifies potential risks to the City's critical infrastructure. The projects presented in the model were compiled using model-based probability functions. The NPCC used IPCC methods to calculate changes in temperature, precipitation, and sea level rise from global climate model simulations based on three greenhouse gas emission scenarios (A1B, A2, B1). The NPCC used 16 GCMs to generate possible changes in temperature and precipitation. Only seven GCMs were used for sea level rise, as sea level rise is not a direct output of most GCMs. The generated sea level rise values for the New York City area include both global and local components.

According to the NPCC report, changes in mean climate and climate extremes could affect many aspects of New York City's water infrastructure. The potential wastewater and drinking water impacts of the projected air temperature change include decreased water quality due to biological and chemical impacts; increased water demand due to a longer growing season; decreased snowpack, which could reduce inflows to reservoirs during the spring thawing season; changes in the ecology of streams due to higher stream temperature, which could limit the amount of water that can be extracted; and increased water demand. The biological and chemical reactions in wastewater treatment plants also could be disrupted at higher temperatures (DEP, 2008a).

Impacts related to the potential changes in precipitation include decreased average reservoir storage, increased turbidity, increased nutrient loads, eutrophication, taste and odor problems, and increased loading of pathogenic bacteria and parasites in reservoirs. Impacts of potential precipitation changes on the wastewater system include increased probability of sewer flooding and increased CSO events.

The potential impacts of sea level rise for city water resources include intrusion of the salt front farther up the Hudson River (NASA, 1999), increased probability of seawater entering sewers, reduced ability of wastewater treatment plants to discharge treated water by gravity alone, increased risk of CSO events, and increased coastal flood risk for low-elevation infrastructure and wastewater treatment plants (NYC Panel on Climate Change, 2009; DEP, 2008a).

3.5. Application of Vulnerability Assessment Information

DEP has conducted a suite of modeling studies to understand the vulnerability of its systems to climate change, and has made decisions to reduce that vulnerability. This situation, however, might be a case where analysis and policy, although informing each other, are proceeding in parallel. New York City's climate change work has led to increased consideration of climate change in strategic planning, has altered operations and maintenance practices, and has changed future infrastructure planning and design. Many of these changes are not the direct result of the New York City vulnerability assessments. Rather, they are part of a larger effort to improve the resiliency and redundancy of water infrastructure in the face of existing vulnerabilities that could be exacerbated by climate change. These decisions largely focus on so-called "no-regrets" opportunities, or changes to the water supply system that make sense regardless of whether or how climate changes. Some of these policy choices have been forced by regulatory mandates, such as the development of a filtration plant for the Croton Watershed, but others have significant benefits system wide, such as reducing leakage from aging supply infrastructure.

PlaNYC and DEP's Climate Change Task Force have identified several initiatives that aim to efficiently and effectively upgrade the city's drinking and wastewater systems in the face of a changing climate. Proposed initiatives are discussed in detail below.

3.5.1. Decreasing Turbidity

Turbidity is a significant drinking water concern in the Catskill and Delaware Water Systems. DEP has addressed this issue historically by adding alum as an "end-of-pipe" solution and engaging in source-water protection measures. Projected increases in intense precipitation events under climate change will most likely increase the turbidity of watersheds beyond historic levels. New York City is continuing its historic programs to address this issue. In the future, DEP will address potential turbidity challenges in the Catskill and Delaware Water Systems by relying more heavily on the soon-to-be-filtered Croton system, a proposed interconnection between the Catskill and Delaware Aqueducts, and operational modifications in how DEP uses the Delaware and Catskill Water Systems during heavy precipitation or turbidity events.

3.5.2. Minimizing Flooding

To minimize urban drainage flooding in New York City during the predicted increased severe weather events, the Climate Change Task Force proposed more frequent cleaning of sewers and maintaining catch basins in flood-prone areas. The task force also promoted green roofs and the reuse of stormwater for "ecologically productive purposes." Green infrastructure has become a significant component of DEP's proposed policies, especially for stormwater management to reduce CSOs (PlaNYC, 2008; NYC Green Infrastructure Plan, 2010).

For example, New York City is planning to expand the Staten Island Bluebelt program, which was created as a natural system to prevent coastal and urban drainage flooding and septic tank failure. It functions by diverting water from wastewater treatment to natural systems. Nearly 36% of Staten Island's precipitation is diverted to a 10,000-acre Bluebelt area. The Bluebelt program has saved the city an estimated $80 million in infrastructure development (Rosenzweig et al., 2007). As severe weather events increase, the Bluebelt and further expansions will act as natural buffers, reducing pressure on the wastewater system and reducing flooding issues and CSOs (PlaNYC, 2008).

3.5.3. Minimizing Supply and Demand Imbalances

Higher air temperatures increase peak water demand. Within New York City, annual average demand is approximately 1,069 mgd. During heat waves, demand can increase to more than 2,000 mgd. To minimize supply and demand imbalances, the Climate Change Task Force has stressed the importance of structural improvements, such as reducing water pressure problems and leakage. Small-scale conservation efforts also can reduce water demand (DEP, 2008a).

New York City has reduced water demand since 1985 through a variety of conservation efforts, including education, metering, water-use regulation, leak detection, installation of magnetic-locking hydrants, and rebate programs. These conservation efforts reduced water consumption from 195 gpd per capita in 1991 to 167 gpd per capita in 1998, with coincident substantial cost savings for both DEP and its customers (U.S. EPA, 2002). Reducing water demand also limits the amount of water entering the wastewater system and, thus, stress on the system. With the above conservation measures, the volume of generated wastewater decreased by 200 mgd between 1996 and 2006 (DEP, 2006).

Additionally, to ensure sufficient water quantity even in the face of higher temperatures, DEP is evaluating new water sources throughout New York City and upstream watersheds. These sources include groundwater and new infrastructure, such as potentially increasing the capacity of the Catskill Aqueduct.

3.5.4. Decreasing Combined Sewer Overflows

To decrease CSO events caused by increased precipitation and intense precipitation events, DEP has begun plans to upgrade wastewater treatment capacity, construct additional holding tanks to increase wet-weather holding capacity, and optimize sewer infrastructure to limit releases. New York City is also planning to convert some of the combined sewer systems into high-level sewer systems,[8] which divert a large percentage of the stormwater directly to waterways rather than into treatment plants. This strategy not only decreases the likelihood of CSO events but also reduces costs by avoiding unnecessary water treatment. The Climate Change Task Force also has proposed increasing pipe size to increase flow in areas where possible (PlaNYC, 2008). In September 2010, DEP released the NYC Green Infrastructure Plan, an adaptive management strategy for reducing CSOs using green infrastructure, grey infrastructure, system optimization, and water conservation.

3.5.5. Adapting to Flood Risk

DEP is also considering converting water storage reservoirs for use as both water supply and riverine flood control (DEP, 2005). To prevent critical assets from being disabled during flood events, the DEP Climate Change Task Force has proposed moving key assets above projected flood heights, installing watertight doors around crucial equipment, switching to submersible pumps, and creating protective barriers such as sea walls, dunes, or tidal gates around important assets (DEP, 2008a).

DEP will institute a snowpack-based reservoir management program to provide enhanced riverine flood attenuation downstream. Under this program, Schoharie Reservoir would be sustained below full capacity during the winter months when sufficient snowpack is present in its watershed so that associated runoff produced by spring snowmelt could refill the reservoir to full storage capacity. The capture of inflows associated with spring storm events and snowmelt runoff in the reservoir would provide additional attenuation in downstream sections. The temporary reservoir level targeted during the snowpack-based reservoir management period would be regularly adjusted based on snow-water equivalent estimates of the watershed's regularly monitored snowpack. As the name implies, snow-water equivalent is the water depth equivalent of a given depth of snow and depends on such factors as the water content and density of the snowpack (DEP, 2008c).

4. Seattle Public Utilities

4.1. Background

Seattle Public Utilities (SPU) was formed in 1997 as a combination of the Drainage and Wastewater Utility, Solid Waste Utility, the former Seattle Water Department, and portions of Seattle City Light and the Seattle Engineering Department. This case study focuses on SPU functions related to supplying drinking water to the City of Seattle, Washington. SPU's drainage and wastewater system are mentioned only briefly.

SPU provides drinking water to a population of more than 1.35 million people in Seattle and surrounding suburban areas. SPU provides direct retail water service to about 630,000 people primarily in the City of Seattle, parts of Shoreline, and small areas just south of the city limits. SPU also sells water wholesale to 25 neighboring cities and water districts serving another 720,000 people. SPU supplied 45.1 billion gallons of drinking water in 2008 from two Cascade Mountain watersheds supplemented with groundwater wells.

The Pacific Northwest climate is dominated by large spatial and temporal variations in precipitation due to maritime influences and extreme topographical variation between the coast and the Cascade Mountains. The low-lying valleys west of the Cascades, including the SPU service area, are characterized by mild temperatures year round, wet winters, and dry summers. Average annual precipitation for the Seattle area is about 37 inches, but in the mountains, that total exceeds 100 inches. About 75% of Seattle area precipitation falls between October and March (Miller and Yates, 2006). The SPU water supply system therefore also must be managed for riverine floods. Typically, early winter precipitation fills reservoirs, which are allowed to spill in anticipation of snowmelt combined with normally rainy springs, which refill reservoirs for the dry summer months.

The regional climate fluctuations known as the El Niño Southern Oscillation (ENSO) and the Pacific Decadal Oscillation (PDO) also strongly affect the Pacific Northwest. ENSO, PDO, and the winter and spring climate of the region are highly correlated, enabling predictions of Pacific Northwest precipitation, snowpack, and streamflow. The University of Washington Climate Impacts Group (UW-CIG) has developed annual climate forecasts for regional resource managers, including annual projections of climate variations due to ENSO and PDO. These forecasts help inform SPU managers about projected conditions over the winter and spring months to enable more informed management of the competing objectives of water supply and flood management (UW-CIG, 2010a).

Observed changes in climate include the following: temperatures increased in the Pacific Northwest by 1.5°F between 1920 and 2003 (Mote, 2003); annual precipitation increased by 14% between 1930 and 1995 (Mote, 2003); April 1 snow-water equivalent has declined dramatically at almost all Pacific Northwest sites (Mote et al., 2003, 2005, 2008; Hamlet et al., 2005); and timing of peak runoff shifted earlier by 0–20 days between 1948 and 2002 (Stewart et al., 2004).

4.2. Description of the Water System

4.2.1. Water Supply System

SPU's water supply comes from three sources: the Cedar River Municipal Watershed, the South Fork Tolt Watershed, and the Seattle Well Fields (Figure 4; SPU, 2010a).

In 1895, Seattle residents voted to approve revenue bonds to construct the Cedar River Municipal Watershed. The watershed, almost entirely owned by the City of Seattle, covers 90,638 acres and provides approximately 70% of the city's freshwater supply over the course of the year. Rain and snowmelt are collected and stored in two reservoirs created by the 1914 construction of the Masonry Dam—Chester Morse Lake and Masonry Pool. As water leaves these reservoirs, it powers the Cedar Falls hydroelectric power plant. Twelve miles downstream, at the Landsburg diversion dam, on average, 22% of the river flow is screened to remove debris, chlorinated for microbial control, and fluoridated for dental health. This water is then stored in Lake Youngs, where it is ozonated for odor and taste improvements, ultraviolet-disinfected to disable chlorine-resistant microbes, chlorinated again, and supplemented with lime for pH-adjusted corrosion control to minimize lead leaching in older plumbing systems.

The Cedar River Municipal Watershed is managed to provide an adequate water supply (both for human use and instream conservation flows). The water supply system also provides riverine flood management and hydropower generation. Morse Lake and Masonry Pool hold, on average, just enough water for one water-cycle year. If too little water is released during winter, flooding can occur during heavy rains or when the snowpack melts during the spring wet season. Winter water levels are therefore generally kept low. Conversely, drought conditions in the spring can prevent the reservoir from refilling to the level necessary to provide water during the dry summer months. This situation presents a risk tradeoff that SPU water managers must address every year to meet both riverine flood management and water supply objectives (SPU, 2010b).

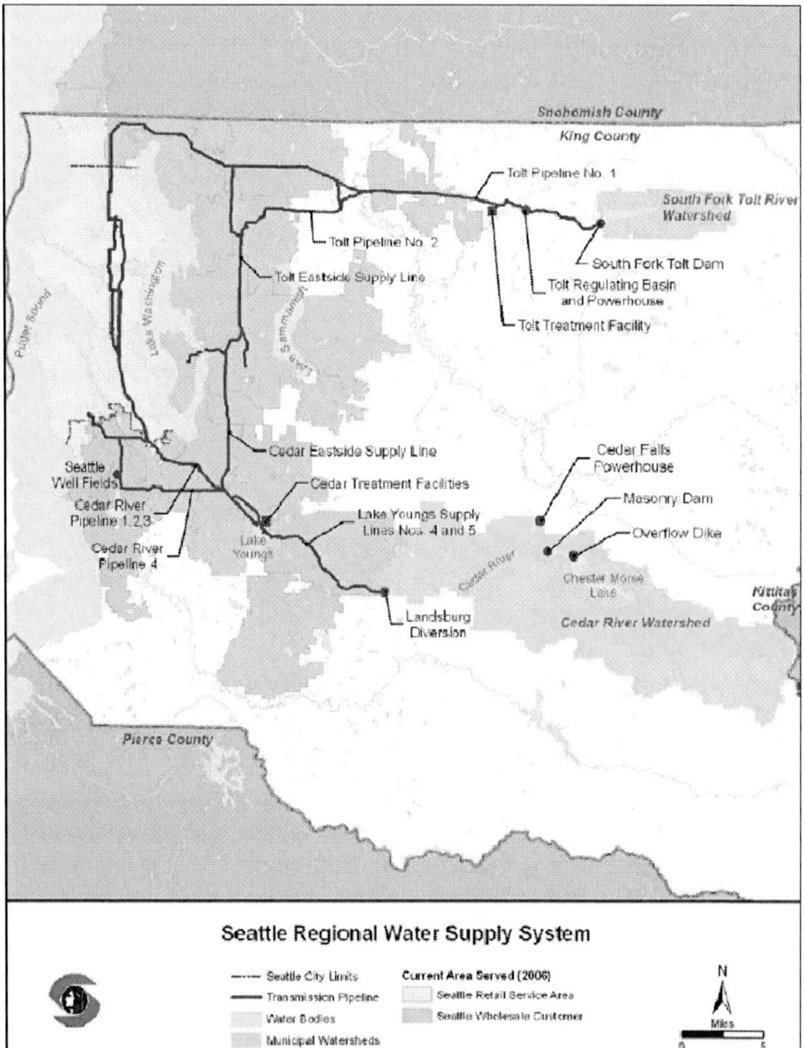

Source: SPU (2008).

Figure 4. Seattle Public Utilities service area.

The South Fork Tolt Watershed is located on the western slope of the Cascade Range, approximately 35 miles east of Seattle (Figure 4). The City of Seattle purchased water rights to the South Fork Tolt from the Mountain Water Company in 1936, but no infrastructure existed for the diversion, conveyance, or distribution of that water. The South Fork Tolt Dam was constructed in 1963, and in 1964, South Fork Tolt Reservoir began supplying water to north Seattle and the Eastside. As water leaves this reservoir, it powers the South Fork Tolt hydroelectric power plant. After a land exchange with Weyerhaeuser Company in 1997, the City of Seattle owned 69% of the 12,107-acre drainage area upstream of the South Fork Tolt Dam. Most of the remaining land lies in the Mt. Baker-Snoqualmie National Forest. The South Fork Tolt Reservoir provides approximately 30% of the city's freshwater supply (SPU, 2008, 2010c). The reservoir is also operated to manage riverine flooding and maintain instream flows.

SPU's Tolt Treatment Facility, the city's first filtration and ozonation facility, began operating in 2000. It provides 120 million gallons per day (mgd) of finished water to customers in Seattle and suburban cities. Although the facility has historically provided very high-quality water, requiring only minimal treatment, it was designed to allow long-range conformity with anticipated regulations, to increase system yield, and to permit continuous supply of Tolt water though periods of high turbidity (SPU, 2010d). Like the Cedar River water supply, the Tolt supply provides fluoridation, chlorination, and adjustment of pH and alkalinity for corrosion control (SPU, 2006a).

In 1987, the first groundwater source was added to the system when two wells in the Highline Well Field began operating. A third well was added in 1990. These wells supply less than 1% of SPU's water from an aquifer to supplement summer demands when necessary. The well field can be pumped for 4 months and becomes available in July (WA DOE, 2001).

Water demand for the SPU system peaked in the 1980s at approximately 170 mgd. A severe drought and mandatory water restrictions in 1992 caused demand to decrease. Subsequently, higher water rates, plumbing code revisions in 1993, conservation efforts, and improved systems operations caused demand to level out around 150 mgd. The economic slowdown in 2000 and continued conservation efforts further reduced demand to approximately 130 mgd. This 24% decrease in demand coincided with a 17% increase in the population of SPU's service area since 1990 (SPU, 2006a). This equates to a 27% decrease in water consumption per capita from 145 gallons per day (gpd) per capita to 105 gpd per capita (SPU, 2010e).

4.2.2. Wastewater System

SPU also conveys wastewater to King County's Wastewater Treatment Division, including associated infrastructure. This drainage infrastructure is partly a combined sewer system, which means that SPU must address the City of Seattle's stormwater quality and urban drainage flooding issues, but often in conjunction with King County Department of Natural Resources and Parks. Because the responsibilities of King County and SPU overlap to some degree, this case study does not present SPU activities related to drainage and wastewater in detail. Of note, however, is that both SPU and King County own conveyance infrastructure within the city, and that combined sewer overflow events are a problem for both entities. SPU has explored the implications of climate change on its stormwater infrastructure and operations. SPU also is pursuing and evaluating adaptation options, engaging in research, and participating in collaborative networks to address stormwater issues—effectively replicating their experience with water supply, but for drainage and wastewater issues.

4.3. Climate Change Projections and Risks

SPU expects climate in the Seattle area to change in several ways. Global climate models (GCMs) project that temperatures will warm at a rate of 0.5°F per decade, nearly three times the rate experienced over the twentieth century. Most models suggest small changes in annual precipitation compared with year-to-year and decade-to-decade variability observed for the twentieth century. Most models, however, do indicate increased winter precipitation and decreased summer precipitation. The potential future impacts of these changes include decreased mountain snowpack, higher winter and lower spring streamflows, increased sea-surface temperatures, rising sea levels, and increased winter riverine flooding (UW-CIG, 2010b; City of Seattle, 2006). Table 3 provides a summary of projected temperature and precipitation changes in the Pacific Northwest from the Washington State Climate Change Impacts Assessment. Projected changes are based on simulations from 20 climate models and two greenhouse gas emissions scenarios (B1 and A1B; UW-CIG, 2010b).

Table 3. Projections of changes in annual mean temperature and precipitation for the 2020s, 2040s, and 2080s

Time period	Temperature °F (*°C*)	Precipitation %
2020s		
Low	+1.1 *(0.6)*	−9
Average	+2.0 *(1.1)*	+1
High	+3.3 *(1.8)*	+12
2040s		
Low	+1.5 *(0.8)*	−11
Average	+3.2 *(1.8)*	+2
High	+5.2 *(2.9)*	+12
2080s		
Low	+2.8 *(1.6)*	−10
Average	+5.3 *(3.0)*	+4
High	+9.7 *(5.4)*	+20

A series of severe droughts in 1987, 1992, and 1997−1998 increased the sensitivity of SPU managers to the effects of climate on their water supply. A very dry summer in 1987 caused significant declines in raw-water supply quality and forced use curtailments, reduced instream flows for fish, and necessitated the installation of an emergency pumping station to access low water in Chester Morse Lake. In response, the City developed a Water Shortage Contingency Plan (updated in SPU, 2006b) and a state-of-the-art reservoir management and streamflow forecasting model for use in real-time water management and long-range planning. The 1992 water shortage was caused by following standard flood control rules after a below-normal winter snowpack. When the spring season also produced below-normal precipitation, SPU's mountain reservoir levels did not recover, and mandatory water restrictions were in place by mid-May. Throughout the summer, raw-water quality declined, leading to a decision to invest in an ozone-purification plant. SPU also implemented dynamic flood-control rules, which, in conjunction with enhanced real-time snow, weather, and streamflow monitoring networks, allowed SPU to implement its new reservoir management approach. Finally, in 1997, new research on ENSO effects on the Pacific Northwest was incorporated into SPU's reservoir management decisions. In anticipation of lower-than-normal snowpack followed by a hot, dry summer, SPU allowed its mountain reservoirs to fill higher than normal and reduced its operational use of water. These proactive decisions allowed the 1997–1998 drought to pass without the public's experiencing any water shortage or restrictions (UW-CIG, 2010c). Another record low snowpack in 2005 threatened water shortages and use restrictions, but, again, careful water management and late spring rains allowed SPU to meet all water supply and instream flow requirements without restrictions.

SPU's predecessor agencies were engaged with the issue of climate change as far back as the 1980s when they helped to develop the American Society of Civil Engineers' policy on global climate change. Informal tracking of climate variability and change issues by SPU staff followed, leading to SPU's formal integration of ENSO into its 1997–1998 reservoir management decisions. In 2002, SPU contracted with UW-CIG to study the potential impacts of climate change and to develop methods for how SPU could incorporate future climate change into its water supply planning (SPU, 2006a). The impacts on water supply projected in the study were reported and incorporated into SPU's 2007 Water System Plan. In a subsequent project, SPU began a new collaboration with UW-CIG to investigate climate change in partnership with the Cascade Water Alliance, Washington State Department of Ecology, and King County Department of Natural Resources and Parks as described below (RWSP, 2010).

Participating agencies formed a Climate Change Technical Committee in spring 2006. The committee included SPU, along with a number of other city, county, state, and tribal government officials managing water resources in the region. The committee met 17 times from March 2006 through December 2007, and drafted a charter in April 2006 containing the following goals:

- Identify the basic building blocks of our understanding of climate change
- Identify what is known about climate change in the Puget Sound region and its potential impacts
- Identify where more information would be useful
- Communicate what is known to other committees in this process
- Document the committee's findings (Palmer, 2007).

The committee reviewed 10 technical reports written by a University of Washington research team before the reports were publicly released. SPU used information from this work to assess impacts to water supply and demand as described in Section 4.4.

SPU also joined several other major utilities to form the Water Utility Climate Alliance (WUCA) in early 2007. WUCA commissioned two climate change white papers; one of which SPU managed. The first white paper outlines potential improvements to scientific models for projecting the impacts of climate change at spatial and temporal scales relevant to utilities (Barsugli et al., 2009). The second white paper outlines decision-making approaches to address climate change in water resource planning and management in the face of uncertainty about future climate conditions (Means et al., 2010). In addition to WUCA, SPU is involved in several other collaborative efforts to enhance the capacity of the water sector to identify and prepare for the impacts of climate change. A staff member from SPU co-chaired the U.S. Environmental Protection Agency's (U.S. EPA's) Climate Ready Water Utilities Working Group, which is developing recommendations on how the U.S. EPA can support a "climate-ready" water sector. SPU is a member of the Water Research Foundation's Climate Change Strategic Initiative Expert Panel, which is assisting the Foundation in developing a multiyear climate change research agenda for the drinking-water sector. This effort is part of a similar one led by the Water Environment Research Foundation to develop climate research for the clean-water sector. Finally, SPU is collaborating in a new project initiated in 2010 called Pilot Utilities Modeling Applications (PUMA). PUMA includes five utilities and four National Oceanic and Atmospheric Administration-funded Regional Integrated Sciences and Assessment (RISA) groups that will be collaborating on downscaling climate models and using them to assess climate change effects on water supply.

SPU operates in a political and managerial environment that supports engagement on climate change adaptation. The City of Seattle took early leadership roles in climate change on both the mitigation and adaptation front. On February 16, 2005, for example, Seattle Mayor Greg Nickels launched the U.S. Conference of Mayors Climate Protection Agreement. Mayor Nickels made climate protection a keystone issue of his administration, creating the City of Seattle's Environmental Action Agenda in 2005, including the Seattle Climate Protection Initiative and the Seattle Climate Action Plan. The need to adapt to changes in water supply was highlighted in the Seattle Climate Action Plan (City of Seattle, 2006). According to this plan, "It is vital that the City—and all levels of government—plan and prepare for the climate change that is inevitable. Because Seattle's water and hydroelectricity are so dependent on the hydrology cycle in the Cascade Mountains, the City has focused its planning and adaptation analysis work there." By the time this was written in 2006, SPU had already begun looking at climate change in earnest.

4.4. Vulnerability Assessment

SPU has commissioned or conducted a series of increasingly sophisticated analyses over the course of many years to examine the vulnerability of their water supply and stormwater infrastructure and operations to climate change. The analyses have benefitted from the expertise of UW climate scientists and others associated with the NOAA RISA program's Climate Impacts Group (UW-CIG). Over the course of nearly a decade, SPU and their collaborators refined a model-driven vulnerability analysis that projects changes in global climate, downscaled those changes to Seattle and its watersheds, and ran those projected changes through SPU's system models to determine how climate change might affect SPU's water supply, water infrastructure, and operations. The SPU study methodology represents a scenario-based approach to vulnerability assessment. For more detailed information on how SPU and other utilities organized their vulnerability assessments, see the companion report, *Climate Change Vulnerability Assessments: A Review of Water Utility Practices* (U.S. EPA, 2010).

4.4.1. Water Supply

Downscaled scenarios of future temperature and precipitation change were developed for the Puget Sound region using three combinations of climate model/greenhouse gas emissions (from the IPCC *Special Report on Emissions Scenarios*; SRES) and four time periods (IPCC, 2000). The three climate model/SRES combinations include a "middle-of-the-road" regional climate change scenario (MPI ECHAM5/A2) with moderate warming and precipitation increase, a significantly wetter and warmer scenario (IPSL-CM4/A2), and a slightly drier and warmer scenario (GIS S-ER/B 1).[9] The four time periods were a 2000 hindcast and projected future conditions for 2025, 2050, and 2075. Climate models were selected because they performed well in other studies replicating temperature and precipitation trends of the Pacific Northwest during the twentieth century (Mote et al., 2005). A statistical downscaling approach was used to translate GCM grid-scale output to a quasi-steady-state daily time series of temperature or precipitation for a specific location at a specific future time that preserves the historic variability of climate (Polebitski et al., 2007a; Polebitski et al., 2007b).

SPU's water supply planning model is the Conjunctive Use Evaluation (CUE) systems model—a weekly time-step simulation model of the Cedar and Tolt River systems. Because it uses observed inflow data for both river systems as input, however, CUE cannot directly incorporate climate model output (temperature and precipitation). Consequently, UW-CIG ran the downscaled meteorological data sets through its Distributed Hydrology Soils and Vegetation Model (DHSVM) to produce climate-altered hydrologic data sets for use in CUE. CUE is used for calculating the firm yield[10] and reliability of Seattle's water supply system and potential future water supply projects. CUE results indicated that yield decreased under all climate change scenarios for all time periods. SPU also ran several planning scenarios through CUE to determine whether available supply could be increased to compensate for anticipated supply shortfalls.

4.4.2. Water Demand

SPU examined the effect of climate change on water demand using a dual approach of regression analysis and forecast modeling. First, SPU performed a regression analysis of peak-season consumption for 1982–2007 using monthly consumption data, maximum temperature, and rainfall at SeaTac Airport for May through September. This relationship was assumed to hold in the future. SPU had also already developed a demand forecasting model for its *2007 Water System Plan* and used this model to forecast non-climate-related changes in demand (SPU, 2006a). In this study, demand was forecast to decrease below historic levels through 2050, but to increase above historic levels by 2075. Applying the results of the regression analysis to these forecasts adjusts demand slightly upward due to the climate change scenarios in 2025 and 2050. In 2075, a larger climate-induced increase in conjunction with significant increases in baseline demand led to 2075 demands increasing to above historic levels.

4.4.3. Storms and Runoff

SPU also engaged a consultant to use the University of Washington's Weather Research and Forecasting (WRF) regional climate model to examine projected precipitation changes in the Thornton Creek Watershed (Northwest Hydraulic Consultants, 2009). This study was focused on storms and runoff, and is distinct from the statistical downscaling study of water supply described previously. Northwest Hydraulic Consultants used the WRF model to dynamically downscale temperature and precipitation data from two GCM-scale simulations (CCSM3 model with A2 SRES, and ECHAM5 model with the A1B SRES) for two 31-year periods (1970–2000 and 2020–2050) to grid sizes of 20 and 36 km^2. These data were used in the rainfall/runoff model Hydrologic Simulation Program-Fortran (HSPF) to model changes in several creek parameters for the entire Thornton Creek Watershed. The results of the study indicated that runoff would increase, except at one sub-basin where modeling results diverged, although the magnitude of the increases varied by a factor of two at times. This work, however, was deemed too uncertain for SPU's planning purposes. According to the study's conclusions, "Additional work is needed to improve confidence in future projections before applying dynamically downscaled data to stormwater planning, policy, or design standards" (Northwest Hydraulic Consultants, 2009). SPU currently is not using modeled climate projections for stormwater planning purposes.

4.5. Application of Vulnerability Assessment Information

SPU arguably has undertaken the most sophisticated vulnerability assessment of any of the utilities discussed in this report, and is the only one of the four utilities that directly used the results to make an adaptation decision. SPU has also identified a more far-reaching set of adaptation options for use in future decades in the event that demand exceeds available water supplies. Prior to this analysis, SPU was already considering how to make the most effective use of usable storage, including the use of dynamic rule curves that use current watershed state conditions instead of relying on past hydrologic records. SPU also had a conservation program that had led to significant reductions in demand since the mid-1980s and more recently, has committed to an additional 15 mgd of conservation by 2030. This analysis demonstrated what SPU already knew, that demand could exceed supplies in 2075 even without climate change, with no new conservation programs past 2030.

4.5.1. Water Supply

Based on the information generated by its water supply vulnerability assessments, SPU determined that available water supply decreased under all scenarios for all time periods. Projections for 2025 indicated Seattle's water supply would decrease by 6–10%, projections for 2050 indicated a decrease of 6–21%, and projections for 2075 indicated a decrease of 13–25%. Demand was projected to decrease in 2025 and 2050 to around 83% of historic supply but increase in 2075 to approximately 106% of historic supply.

Based on these projections, SPU analyzed "Tier 1" low- or no-cost intrasystem modifications that effectively increased the usable storage capacity for water with no new supply infrastructure.[11] This approach primarily consisted of eliminating conservative operating assumptions from SPU's water system supply calculations. [12] These low-cost modifications were estimated to compensate for supply shortfalls in all three scenarios in 2025, in two of three scenarios in 2050, and in none of the three scenarios in 2075.

"Tier 2" alternatives were identified that could compensate for the remaining projected shortfalls in 2050 and 2075. These included additional use of Lake Youngs storage, modified/optimized conjunctive use operations, and additional conservation programs after 2030. Even more expensive or complex alternatives were identified for "Tier 3," "Tier 4," and "Tier 5" spanning from reservoir operational changes to new supply alternatives, but these higher cost modifications were deemed unnecessary through 2075.

All of these policy options were directly informed by the quantitative results of the SPU vulnerability analysis. SPU has decided that its vulnerability analysis indicates no need for near-term operational changes or new infrastructure. In one sense, SPU has not changed its water supply planning and management decisions because, although significant, the projected climate impacts are not imminent. On the other hand, the changes to conservative operating assumptions represent an important class of "no-regrets" adaptations. Even the Tier 2 adaptations, such as increased water conservation efforts, represent policy options that provide benefits in terms of supply reliability, regardless of the magnitude of climate change. In SPU's current estimation, no adaptations beyond Tier 2 will be needed through 2075.

4.5.2. Storms and Runoff

The results of SPU's dynamical downscaling and urban drainage study provided insufficient certainty to be useful for planning purposes. Consequently, SPU is relying on a qualitative understanding that intense precipitation events are likely to increase and is exploring changes in peak storm events as a proxy for changes in precipitation due to climate change. This approach represents an important hedging strategy. In the absence of reliable projections of future climate conditions, SPU decided to apply a safety factor to new infrastructure construction to ensure that new investments would perform their intended function over their useful lives based on a general understanding of the climate trends and a reasonable estimate of the magnitude of that change.

5. SPARTANBURG WATER

Spartanburg Water is a public water and wastewater utility that is composed of two distinct legal entities: Spartanburg Water System (SWS) and

Spartanburg Sanitary Sewer District (SSSD). The two entities function as one company (West, 2010). SSSD was formed as a special-purpose district for wastewater services. SWS is a political subdivision of the City of Spartanburg and is overseen by three Commissioners of Public Works. SSSD is overseen by the seven-member Sewer Commission, which includes the three Commissioners of Public Works (Spartanburg Water, 2010a; West, 2010).

5.1. Background

Spartanburg Water serves a population of approximately 180,000 people in Spartanburg County and portions of Greenville, Cherokee, and Union Counties in South Carolina (Spartanburg Water, 2010a). The SWS service area includes a contiguous retail service area of approximately 259 mi^2, a noncontiguous retail service area of approximately 15 mi^2, and a wholesale service area of approximately 605 mi^2(Spartanburg Water, 2010a). The SSSD wastewater service area is defined by the Spartanburg city limits—a contiguous service area covering approximately 196 mi^2, and a noncontiguous service area of eight locations serving approximately 22 mi^2 (Spartanburg Water, 2010a).

From 1971 to 2000, Spartanburg County received on average 61 inches of rain per year (SRCC, 2010). Precipitation is somewhat consistent throughout the year, ranging on average from 3.44 to 6.86 inches per month (SRCC, 2010). The average minimum and maximum temperatures are 48.6°F and 71.3°F, respectively (SRCC, 2010). Over the past 10 years, however, the southeastern region of the United States has experienced prolonged droughts that have lasted several years. The Spartanburg Water service area experienced droughts in 2002 and 2003, and a drought has persisted since 2005, with the lowest recorded streamflow occurring in 2009 (West, 2010).

5.2. Description of the Water System

5.2.1. Water Supply System

Spartanburg Water provides approximately 30 million gallons of water to its customers each day. Approximately 60% of the water use is residential (West, 2010). Industrial water use has significantly declined in the past decade from 110 to 52 industrial accounts (West, 2010). Although there are commercial and other small business accounts, these sectors are not significant water users.

Three reservoirs on the Pacolet River provide the vast majority of the Spartanburg water supply (Figure 5). Bowen Reservoir, the most northern reservoir, is on the south fork of the Pacolet River. Built in 1960, it covers 1,534 acres and has a total capacity of 17,115 acre-feet (Spartanburg Water, 20 10b; West, 2010). Water from Bowen Reservoir flows downstream to Municipal Reservoir Number 1 (MR1), which is located just above the confluence of the North and South Pacolet Rivers. Built in 1926, MR1 serves mainly as a pass-through reservoir with a capacity to store approximately 1 day's worth of water. MR1 improves water quality through sedimentation as the water flows through it (West, 2010). Blalock Reservoir is downstream from the confluence of the North and South Pacolet Rivers and receives inflow from MR1 and the North Pacolet River. Built in 1983, Blalock Reservoir covers 1,105 acres and has a total capacity of 16,894 acre-feet (Spartanburg Water, 2010b; West, 2010). Spartanburg Water expanded Blalock Reservoir in 2006 by raising the height of dam to meet projected future water demand (Spartanburg Water, 2009).

Two smaller water sources supplement the reservoir system: Vaughn Creek and an unnamed stream off Hogback Mountain provide approximately 800,000–900,000 gallons per day (gpd) of water. These sources supply the Landrum Water Treatment Plant (WTP), which has a capacity of 1 million gallons per day (mgd). Landrum WTP also has a groundwater backup source with a pumping capacity of 50,000 gpd (West, 2010).

In addition to Landrum WTP, two other plants provide drinking water treatment. R.B. Simms WTP, located at Bowen Reservoir, has a capacity of 64 mgd, and Blalock WTP, located at Blalock Reservoir, has a capacity of 22.5 mgd. Spartanburg's three water treatment plants provide full conventional treatment, including sedimentation, filtration, and chlorination (West, 2010).

The distribution system is composed of 1,308 miles of pipes (West, 2010). Hydroelectricity produced at R.B. Simms WTP is used to support the water treatment operations. Hydroelectricity is not generated, however, when Spartanburg Water operates in full conservation mode during droughts. Operating in conservation mode can have a significant effect on the utility's energy costs, especially during peak hours, because peak usage can set the pricing for the month for all of its electricity use (West, 2010).

Discharges from Blalock Reservoir are managed to meet instream flow requirements based on a combination of factors, including the water level in Blalock Reservoir, the time of year, and flow into the reservoir system.

Because of spawning of fish downstream, the South Carolina Department of Health and Environmental Control (SC DHEC) issued Spartanburg Water a permit that set the downstream flow requirements and determined the 7Q10 for Pacolet River[13]. In the event of a persistent drought, Spartanburg Water may request permission for reduced releases, provided it conducts additional monitoring to ensure fish health and water quality (West, 2010).

5.2.2. Wastewater System

Spartanburg Water has 10 wastewater treatment plants (WWTPs) that range in capacity from 50,000 gpd to 25 mgd. The largest of the 10 plants, Fairforest WWTP, is located just downstream of Blalock Reservoir. All 10 WWTPs provide secondary treatment. Discharge permits for the WWTPs are calculated based on the 7Q10, which is determined in part by releases from the reservoirs; therefore, there is a relationship between Spartanburg Water's ability to withdraw water and discharge wastewater. In total, approximately 13 mgd of wastewater is collected, treated, and discharged into the Pacolet River (West, 2010).

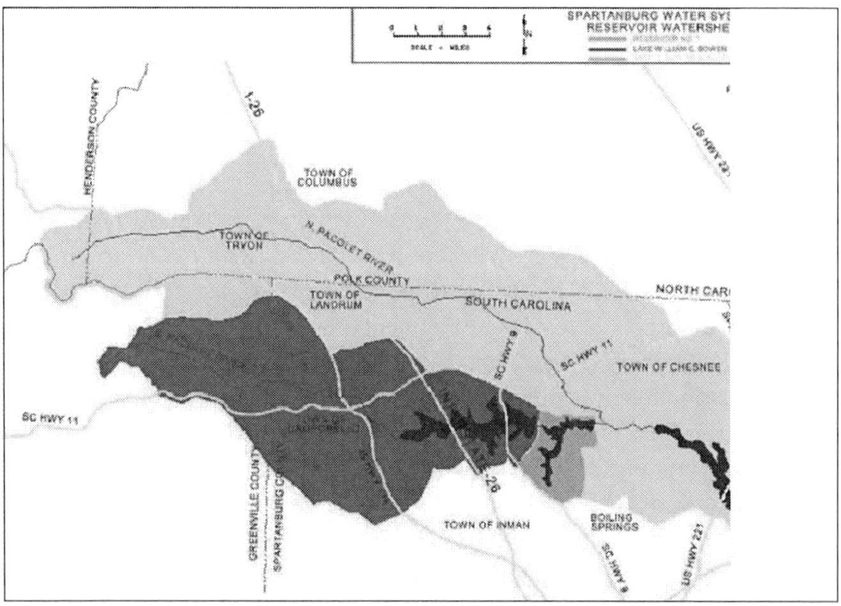

Source: Spartanburg Water (2009).

Figure 5. Reservoirs and watersheds of the Spartanburg water system.

Spartanburg Water owns, operates, and maintains 940 miles of wastewater collection pipes throughout the service area (West, 2010). Some infiltration and inflow occur during storm events. A separate storm sewer system is maintained by the city and county and does not fall under Spartanburg Water's jurisdiction (West, 2010).

5.3. Climate Change Projections and Risks

Spartanburg Water is aware that projected climate change could affect its water and wastewater systems. The utility expects droughts in the region to increase in frequency and severity with greater variability in precipitation. Spartanburg Water also expects severe storm events, such as hurricanes and tropical storms, to increase in frequency and severity.

Spartanburg Water identified these projected climate changes and their potential effects on its water and wastewater systems and operations in a variety of ways. One staff member attends meetings of the South Carolina State Drought Response Committee, whose chair is the State Climatologist and where projected climate change for the area is often discussed (Spartanburg Water, 2010c). Networking within the water utility community has provided information on approaches other utilities are using to examine and address climate change. For example, Spartanburg Water's Deputy General Manager of Engineering and Technical Services served as the President of Water Environment Federation. This provided her the opportunity to visit other domestic water utilities, including those in Las Vegas, East Bay in California, and Seattle, in addition to water utilities in Europe, South Africa, and Tanzania. Also, she and other Spartanburg Water staff often attend conferences to follow the activities of and consult with other utilities (West, 2010).

Mainly because of the prolonged and repeated droughts in recent years, Spartanburg Water is considering the effects of climate change on its infrastructure and operations. The droughts of the past decade have exacerbated many of the existing vulnerabilities of the Spartanburg Water system, including increased water demand from population growth, changes in land use patterns affecting water quantity and quality, and increasing frequency of droughts and extreme storm events affecting quality and flooding (West, 2010).

Most of the expected effects of climate change will require increased management of existing vulnerabilities, rather than addressing completely new challenges (West, 2010). For example, one of Spartanburg's water conservation efforts for drought management is a pricing structure with increasing block rates, which discourages water use beyond a certain minimum level and generally serves to discourage outdoor water use. With potential for more frequent or more severe droughts with climate change, Spartanburg Water might implement this pricing structure along with other enhanced drought management approaches to conserve additional water.

5.4. Climate Change Vulnerability Assessments

Spartanburg Water assessed their potential vulnerability to climate change using existing information on climate change projections, experience from recent extreme events in their region, and experiences of other utilities. Based on this assessment, Spartanburg Water believes the effects of climate change will exacerbate existing vulnerabilities. As a result, rather than undertaking completely new activities or management approaches, the utility is incorporating climate change into many of its existing management activities. Climate change is now a consideration in all utility planning processes and incorporating climate change is part of the utility's culture (West, 2010). To better consider climate change in its decisions, Spartanburg Water attempts to stay current on regional climate change projection data. It also collects and tracks a variety of data relevant to climate change, including rainfall, temperature, streamflow, reservoir levels, groundwater levels, water usage, revenue streams, public perception, and Web sites (West, 2010).

Spartanburg Water's consideration of climate change takes into account the potential effects of climate change throughout its entire system—from providing sufficient water supplies to ensuring an uninterrupted supply chain for treatment chemicals during extreme weather events (West, 2010). This holistic approach to the system and operations—the result of Spartanburg Water's past experiences (such as having an interrupted supply of treatment chemicals following Hurricane Katrina) or lessons learned from other utilities—is essential because many aspects of the system are interconnected. For example, the release of water from Blalock Reservoir for the water system affects the 7Q10 determination for wastewater discharge permitting. Another example of Spartanburg's system-wide thinking is its understanding of the potential effect of water conservation programs on its revenue.

Spartanburg Water has a reservoir management model to support its management and water use decisions throughout the supply system. It also is currently developing a hydraulic model for the wastewater system. Combined with information on projected climate change, Spartanburg Water believes these models of their existing systems will enable them to assess the potential consequences of climate change on the system and allow the utility to consider adaptation actions accordingly.

5.4.1. Water Quantity

One of the primary considerations for the water supply system is sufficient water quantity. In the past 10 years, population growth has increased water demand in the Spartanburg Water service area and in six other water districts in the county downstream of Spartanburg Water. Additionally, continued development has led to more impervious surfaces, which in some cases have redirected runoff outside of the reservoirs' watersheds, thereby reducing supply.

The recent prolonged drought experienced in the region has affected not only surface water but groundwater resources as well. Although the main water source for the Spartanburg Water system is surface water, groundwater contributes to baseflow and, therefore, surface water supplies. Multiple and prolonged years of drought impact groundwater supplies, which can take several years to recover. This results in a continued risk to surface water supply sources beyond the length of the drought (West, 2010). Also, during these drought periods, people living within Spartanburg County who obtain their drinking water from groundwater sources and are not serviced by Spartanburg Water request to be added as a customer because groundwater sources are insufficient. Often these requests originate from areas where water distribution systems do not already exist. This issue can be a challenging to manage because often public expectations are for the utility to provide access to water (West, 2010).

Change in water quantity can also affect wastewater system operations. Several of Spartanburg Water's WWTPs discharge to small streams, where wastewater discharges can constitute up to 80% of streamflow (West, 2010). With prolonged drought, Spartanburg Water anticipates that the future permit limits for these facilities will change if the 7Q10 changes for the receiving streams. In an adjacent county, similar conditions resulted in the wastewater utility upgrading to tertiary treatment. Some of the 7Q10 determinations are expected to undergo review in 2012. The result of these reviews could require additional capital planning and "creative treatment strategies in the interim" (West, 2010).

5.4.2. Water Quality

In addition to drought conditions, the Spartanburg Water service area has experienced extreme rain events, including tropical storms and hurricanes. These events caused flooding throughout the area and damaged components of Spartanburg Water's facilities (West, 2010). With the increased frequency and severity of such storms projected to happen because of climate change, preventive and restorative efforts will require additional planning and financing. Additionally, when combined with continued land development, water quality problems resulting from impervious surface runoff will be exacerbated. Impervious surfaces are a significant water quality concern because runoff of sediments, contaminants, oil, and grease is increased. Because Spartanburg County has no zoning laws, land use changes (and accompanying development and impervious surface increases) can be unpredictable in some areas (West, 2010).

Extreme storm events and droughts, especially in combination, have been associated with taste and odor problems in Bowen Reservoir and MR1. The problems are caused by high levels of geosim, which is a naturally occurring compound produced by certain soil bacteria and blue-green algae, depending on environmental conditions, including water temperatures, nutrient enrichment, and turbidity (USGS, 2009). SC DHEC determined that both reservoirs were fully supportive of all uses based on established criteria. An investigation by Spartanburg Water and the U.S. Geological Survey found that nutrient enrichment was not the contributing cause of the elevated geosim levels but that streamflow and the resulting hydraulics within the reservoir system affect the production and release of geosim by blue-green algae (West, 2010). Hydraulics affects sedimentation or re-suspension of sediment, which, in turn, affects the penetration of sunlight and the temperature of the water—both factors in the release of geosim. Today, Spartanburg Water has a monitoring system in place that helps predict when geosim events might occur.

The two main weather events that can trigger geosim release are (1) droughts, when water clarity is greatly increased by lower reservoir levels and slower streamflow; and (2) major storm events, when hydraulic surges stir up sediment in the system, releasing phosphorus and resulting in an increased abundance of blue-green algae. The highest levels of geosim were observed in 2003–2005 when tropical storms followed droughts in 2002 and 2003(West,

2010). Because these two main contributing factors are both predicted to increase in severity and frequency, Spartanburg Water expects that climate change will exacerbate the geosim water quality problem and could require additional management.

5.4.3. Infiltration/Inflow

With projected increases in the intensity of storm events, Spartanburg expects infiltration and inflow into the wastewater collection system to increase. Increased infiltration and inflow could threaten the capacity of the system to handle wastewater flow during these events.

5.5. Application of Vulnerability Assessment Information

Spartanburg Water does not expect climate change to introduce new challenges but rather to exacerbate existing vulnerabilities. Given both its climate experiences of the past 10 years in the form of increased frequency and duration of drought and its information about projected climate change, Spartanburg Water has initiated a utility-wide effort to incorporate climate change into its planning processes. Spartanburg Water has combined its environmental, operational, and financial data with its understanding of the water system to qualitatively assess the potential effects of projected climate change on its infrastructure, operations, and customer needs. It has also changed its planning and management, particularly by increasing the flexibility of its system and operations.

As part of long-term water supply planning, in 2006, Spartanburg Water doubled the capacity of Blalock Reservoir, based on a study conducted in the 1990s. When this project was completed, Spartanburg Water rethought its management strategy for the reservoir system as a result of extended droughts in the Southeast. Based on a set of indicators, including streamflow and long-term weather forecasts, reservoir releases are managed for maximum water storage when signs of prolonged drought are present and hydroelectric power generation might be suspended[14]. Also, because Bowen and Blalock Reservoirs support recreational activities and adjoining properties are permitted to directly withdraw water from the reservoirs for lawn irrigation, recreational activities were limited, and water conservation requirements were instituted. Spartanburg Water asserted its right to discontinue all withdrawals for lawn irrigation from the reservoirs during droughts (West, 2010).

Spartanburg Water also revised its Water Demand Management Plans and became a WaterSense® Partner and Charter Sponsor of The Alliance for Water Efficiency. Aggressive water conservation campaigns were launched throughout the community, including placing educational kiosks and rotating them through public areas. Spartanburg Water promotes water conservation year round, regardless of drought conditions.

As a result of the conservation program and decreased industrial water use, Spartanburg Water has realized a reduction of 10 mgd in water use, and the summer peak demand has been reduced by 5 mgd. Because of this decreased demand, the time that water resides in the distribution system has increased, so Spartanburg Water is considering taking some ground storage offline or retrofitting lines to minimize this time or both. These lines and storage options, however, will be maintained in place for future use. This plan could prove useful, with potential increases in water demand from new customers or increased demand with climate change. In the next 4 years, Spartanburg Water plans to spend $3 million on its water distribution system (Spartanburg Water, 2010d).

The combination of Spartanburg Water's water conservation program and loss of many industrial accounts has resulted in a sustained loss of approximately 13% of the utility's typical revenues in the past 2 years. Spartanburg Water is now re-evaluating its revenue streams and management strategies to ensure not only environmental but also financial sustainability (West, 2010).

On the wastewater side of the system, Spartanburg Water has plans to evaluate the feasibility of modifying future treatment at 3 of the 10 WWTPs (Spartanburg Water, 2010e). The three benefits cited for the projects include the assurance of the effectiveness of ultraviolet light disinfection at these plants, potential future reuse of water, and continued compliance with discharge permits by providing "additional treatment that may be needed with fluctuations in stream flows from climate impacts" (Spartanburg Water, 2010e).

To address the potential increase in infiltration and inflow into the wastewater collection system, Spartanburg Water adopted a new strategy when upgrading pipes in the wastewater collection system. Instead of closing off the old pipes, they were left in place to provide additional capacity during storm events and additional flexibility in managing the projected effects of climate change.

Spartanburg Water has been able to gather information on climate change impacts and adaptation by attending state-level drought committee meetings and networking with other water utilities. It does not, however, have the benefit of a state-level water program that assesses available water resources or provides guidance on projected climate change. Not having a state-level water program limits Spartanburg Water's long-term planning and modeling. Spartanburg Water notes that the most helpful resources for continuing to address climate change would be the availability of more environmental data, descriptions of best practices and management tools, and case studies of other utilities' actions. Addressing overlapping regulations and coordinating regulations on a watershed basis among surface water, groundwater, waste-water, stormwater, and other water resource-related programs also would facilitate Spartanburg Water's efforts to manage its water resources efficiently and sustainably in the face of continued climate change (West, 2010).

6. SUMMARY

Each utility featured in this report is faced with a unique set of issues and challenges related to climate change. Although the issues and challenges vary, several summary observations can be made that might be useful to other utilities and members of the water resources community regarding the conduct and use of climate change vulnerability assessments to support adaptation.

- *Conducting climate change vulnerability assessments appears to have increased awareness of climate change risks, informed decision making, and supported adaptation at the utilities featured in this report.* One theme emerging from the case studies is the need to consider climate change in a holistic context, taking into account all factors affecting system performance.
- *Utilities have worked with climate scientists and modelers to obtain data and gain insight into how climate science can be used to inform their decision making.* SPU collaborated with the UW-CIG, DEP collaborated with Columbia University and the City University of New York, and EBMUD used an analysis conducted by the State of California and the California Climate Change Center. In contrast, Spartanburg relied on information gathered from briefings and staff contact with other utilities through participation in the Water Environment Federation and the American Water Works Association but did not formally collaborate with the climate change research

community to develop region-specific data on their climate change risks.

- *Uncertainties in vulnerability assessments or climate science need not delay adaptation action.* All four utilities described in this report have taken action to reduce their climate risk despite significant, remaining uncertainties regarding the potential impacts of climate change. Often, these actions also address other concerns, such as managing limited supplies using water conservation or water reuse. Nonetheless, climate change vulnerability analyses can help inform and prioritize among management decisions.
- *The large utilities used a wide array of climate change scenarios to capture some of the uncertainty about future climate change.* EBMUD conducted a "bottom-up" approach by performing sensitivity analyses to improve its understanding about how climate change could affect particular elements of its water resource system. SPU and DEP conducted what are often referred to as "top-down" approaches driven by climate change scenarios and models.[15]
- *Vulnerability analyses to date have focused mainly on water supply and demand.* All four utilities focused their climate change impacts and vulnerability work principally on water supply and demand. Although the utilities expressed concern about the effects of climate change on water quality, urban drainage, wastewater, and other aspects of their systems, these areas have not yet received the same level of attention as water supply.
- *The utilities used system-specific models to understand and manage potential climate impacts on their systems.* All case studies except Spartanburg Water used their system hydrology, operations, or planning models to evaluate the effects of potential climate change on their systems. The models were used to assess whether operational changes would be sufficient to cope with the effects of climate change, or whether system changes, such as adding supplies or further reducing demand, also were necessary. These models did not always align with the output of climate models, necessitating tailored use of climate projections for each utility. Spartanburg Water used existing system models to understand their system's behavior and qualitatively determine expected climate change impacts.
- *The case studies demonstrate that a variety of methods can be used to understand vulnerability and analyze adaptation options.* Scenario analyses, sensitivity analyses, state-of-the-science literature reviews,

and peer information sharing were used in different combinations in the four case studies to understand the potential impacts of climate change. The sensitivity analysis EBMUD performed on their existing system model and the literature review and information gathering from peers Spartanburg Water performed demonstrate two paths that utilities lacking the financial and staff resources to support detailed modeling studies can take to assess their vulnerability to climate change.

- *Utilities expressed an interest in having their needs reflected in future research.* Utilities specifically requested higher resolution climate change projections at the spatial and temporal scales at which they operate, probabilities associated with projected changes, and guidance on appropriate climate change parameters and scenarios to consider and plan for in their regions. One recommendation was to establish a central repository of data to support climate change and adaptation analysis.
- *The results of vulnerability assessments by the four utilities were used in different ways to inform and support adaptation.* SPU responded directly to the results of the vulnerability analysis by evaluating the impact of conservative operational assumptions on reservoir management. The other utilities used their vulnerability assessments to increase knowledge about their climate change risks, integrate information on these risks into decision making, and provide support for adaptation measures.

REFERENCES

[43] AWWA (American Water Works Association). (1997). Climate change and water resources. Committee report of the AWWA public advisory forum. *J Am Water Works Assoc*, *89*(11), 107−1 10.

[44] Barsugli, J., Anderson, C., Smith, J.B. et al. (2009). *Options for improving climate modeling to assist water utility planning for climate change.* Prepared for Water Utility Climate Alliance. December. Available online at http://www.wucaonline.org/assets/pdf/actions whitepaper 120909.pdf.

[45] Brunette, V. & Germain, R. (2003). *Forest management in the New York City watershed.* World Forestry Conference, Quebec City. Available online at http://www.fao.org/docrep/article/wfc/xii/0649-b3.htm.

[46] Chan, C. (2010). Personal communication on April 2, 2010, between Clifford Chan, Manager of Water Treatment and Distribution, EBMUD, and Viktoria Zoltay, Abt Associates.

[47] Cheng, S. (2010). Personal communication on April 5 and 7, 2010, between Stephanie Cheng, Associate Engineer in Resource Recovery Program, EBMUD, and Viktoria Zoltay, Abt Associates.

[48] City of Seattle. (2006). Seattle, a climate of change: meeting the Kyoto challenge. Climate Action Plan. September. Available online at http://www.cityofseattle.net/ climate/docs/SeaCAPplan.pdf.

[49] DEP (New York City Department of Environmental Protection). (2005) *Adapting New York City's water supply and wastewater treatment center to climate change*. U.S. Climate Change Science Workshop. November 15. Available online at http://www.climatescience.gov/workshop2005/ presentations/WA2.5_Lloyd.pdf.

[50] DEP (New York City Department of Environmental Protection). (2006). *New York City's wastewater treatment system*. Available online at http://www.nyc.gov/html/ dep/pdf/wwsystem.pdf.

[51] DEP (New York City Department of Environmental Protection). (2008a). *Assessment and action plan: a report based on the ongoing work of the DEP climate change task force*. Climate change program. Available online at http://www.nyc.gov/html/dep/pdf/climate/climate complete.pdf.

[52] DEP (New York City Department of Environmental Protection). (2008b). *New York City 2008 drinking water supply and quality report*. Available online at http://www.nyc.gov/html/dep/pdf/wsstate08.pdf.

[53] DEP (New York City Department of Environmental Protection). (2008c) Environmental assessment: attachment 1 -- project description. Available online at http://www.nyc.gov/ html/dep/pdf/ gilboa/ gilboa_ proj_desc.pdf.

[54] DEP (New York City Department of Environmental Protection). (2010a). *History of New York City's water supply system*. Available online at http://www.nyc.gov/html/ dep/html/ drinking_water/ history. shtml.

[55] DEP (New York City Department of Environmental Protection). (2010b). *The Croton water filtration plant project*. Available online at http://www.nyc.gov/html/ dep/html/news/croton.shtml.

[56] Dettinger, M.D. (2004). *From climate-change spaghetti to climate-change distribution. In: Climate scenarios and projections: the known, the unknown and the unknowable as applied to California*. Aspen Global

Change Institute Workshop held March 11−14, 2004, San Francisco, CA. Available online at http://www.agci.org/programs/past_ scientist_ workshops/about_the_workshop/sciSess_details.php?recordID=88 (as cited in EBMUD 2009a).

[57] EBMUD (East Bay Municipal Utility District). (2007). *All about EBMUD*. Available online at http://ebmud.com/sites/default/files/pdfs/ current allaboutebmud[1].pdf.

[58] EBMUD (East Bay Municipal Utility District). (2008). *Strategic plan.* July. Available online at http://ebmud.com/sites/default/files/pdfs/ Strategic_ Plan_July_2008[1].pdf.

[59] EBMUD (East Bay Municipal Utility District). (2009a). *Climate change analysis*. In: WSMP 2040—water supply management program 2040 plan, Appendix C TM-9, October. Available online at http://www. ebmud.com/sites/default/files/pdfs/WSMP-2040-Appendices-October-2009.pdf.

[60] EBMUD (East Bay Municipal Utility District). (2009b). *WSMP 2040—water supply management program 2040 plan.* Final program environmental impact report response to comments. SCH #2008052006. October. Available online at http://www.ebmud.com/ sites/ default/ files/pdfs/WSMP-2040-Main-Document-October-2009.pdf.

[61] EBMUD (East Bay Municipal Utility District). (2009c). *Water visions: a clear path to the future.* Available online at http://www.ebmud.com/ sites/ default/files/pdfs/ FY09annual10_0.pdf.

[62] EBMUD (East Bay Municipal Utility District). (2010). *Sewer system management plan. Part E8—capacity management.* Available online at http://ebmud.com/sites/default/files/pdfs/New_Construction/E8-Capacity Management.pdf.

[63] Freas, K., Bailey, B., Munevar, A. et al. (2008). Incorporating climate change in water planning. *Journal of the American Water Works Association*, *100*, 6−98.

[64] Hamlet, A.F., Mote, P.W., Clark, M. et al. (2005). Effects of temperature and precipitation variability on snowpack trends in the western United States. *J Climate*, *18*(21), 4545−4561.

[65] IPCC (Intergovernmental Panel on Climate Change). (2000). Emissions scenarios: summary for policymakers. A special report of IPCC Working Group III of the Intergovernmental Panel on Climate Change. Available online at www.ipcc.ch/pdf/special-reports/spm/sres-en.pdf.

[66] IPCC (Intergovernmental Panel on Climate Change), (2007). Climate Change 2007: impacts, adaptation, and vulnerability. Contribution of

Working Group II to the Fourth Assessment Report of the Intergovernmental Panel on Climate Change, eds., Parry, M; Canziani, O; Palutikof, J; et al. Cambridge University Press, Cambridge, United Kingdom and New York, NY, USA.

[67] IPCC (Intergovernmental Panel on Climate Change). (2008). *Climate change and water: IPCC Technical Paper VI.* Available online at http://www.ipcc.ch/pdf/technical-papers/climate-change-water-en.pdf.

[68] LAFCO (Local Agency Formation Commission). (2008). *Water and wastewater services: municipal services review for west Contra Costa County*. August. Available online at http://www.contracostalafco.org/ municipal_service_reviews/west_county_water_wastewater/ 9.0%20EBMUD%20F inal.pdf.

[69] Major, D.C., Horton, R. & Rosenzweig, C. (2007). *Integrated modeling project for water quantity and quality: guidelines for project final report.* Climate impact assessment of environmental infrastructure systems. New York City Department of Environmental Protection. June 29.

[70] Maurer, P., Brekke, L., Pruitt, Y. et al. (2007). Fine-resolution climate projections enhance regional climate change impact studies. *Eos Trans AGU*, *88*(47), 504.

[71] McCarthy, J.J., Canziani, O.V., Leary, N.A. et al, eds. (2001). *Climate change 2001: impacts, adaptation, and vulnerability*. Contribution of Working Group II to the third assessment report of the Intergovernmental Panel on Climate Change. Cambridge, UK, and New York, NY: Cambridge University Press. Available online at http://www.grida.no/publications/other/ipcc tar/.

[72] Means III, E.G., Laugier, M.C. & Daw, J.A. (2010). Decision support planning methods: incorporating climate change uncertainties into water planning. Final report. Prepared for Water Utility Climate Alliance. January. Available online at http://www.wucaonline.org/assets/ pdf/actions whitepaper 0121 10.pdf.

[73] Miller, K. & Yates, D. (2006). Climate change and water resources: a primer for municipal water providers. Denver, CO: Water Research Foundation and University Corporation for Atmospheric Research. Available online at http://www.waterresearchfoundation.org/ research/ TopicsAndProjects/projectProfile.aspx?pn=2973.

[74] Mote, P.W. (2003). Trends in temperature and precipitation in the Pacific Northwest during the twentieth century. *Northwest Sci*, *77*(4), 271–282.

[75] Mote, P.W. Parson, E.A. Hamlet, A.F. et al. (2003). Preparing for climatic change: the water, salmon, and forests of the Pacific Northwest. *Climatic Change*, *61*(1–2), 45–88.

[76] Mote, P.W., Hamlet, A.F., Clark, M. et al. (2005) Declining mountain snowpack in western North America. *Bull Am Meteorol Soc*, *86*(1), 39–49. Available online at http://www.livingrivers.org/pdfs/LRlibrary/ClimateDocs/MoteHamletClarkLettenmaier.pdf .

[77] Mote, P.W., Hamlet, A.F. & Salathé, E.P. (2008). Has spring snowpack declined in the Washington Cascades? *Hydrol Earth Syst Sci*, *12*, 193–206. Available online at www.hydrol-earth-syst-sci.net/12/.../ hess-12-1 93-2008.html.

[78] NASA (National Aeronautics and Atmospheric Administration). (1999). Scientists prepare New York City for future climate change. Goddard Institute for Space Studies. Research news June 4, 1999. Available online at http://www.giss.nasa.gov/research/news/19990604/.

[79] NCDC (National Climatic Data Center). (2010). Climate of California. Climates of the States, Climatography of the United States No. 60. Available online at http://cdo.ncdc.noaa.gov/ climatenormals/clim60/ states/Clim CA 01.pdf.

[80] New York State Climate Office. (2010). The climate of New York. Available online at http://nysc.eas.cornell.edu/climate_of_ny.html.

[81] New York Water. (2010). The threat to NYC's drinking water from phosphorus pollution. Available online at http://www.newyorkwater.org/pdf/phosporousPollutionmar2010 2.pdf.

[82] Northwest Hydraulic Consultants. (2009). *Application of dynamically downscaled climate model data to analysis of hydrologic change in Seattle's Urban Creeks: The Thornton Creek Case Study*. Prepared for Seattle Public Utilities.

[83] NYC (New York City) Green Infrastructure Plan: A Sustainable Strategy for Clean Waterways, (2010). Available online at: http://www.nyc.gov/html/dep/pdf/green_infrastructure/NYCGreenInfrastructurePlan_LowRes.pdf

[84] NYC (New York City) Panel on Climate Change. (2009). Climate risk information. Available online at http://www.nyc.gov/html/ om/ pdf/2009/NPCC CRI.pdf.

[85] NYSDEC (New York State Department of Environmental Conservation). (2010). State enforcement targets quality of water near New York City. Environment DEC. Available online at http://www.dec.ny.gov/environmentdec/18679.html.

[86] Palmer, R.N. (2007). Final report of the climate change technical committee. A report prepared by the Climate Change Technical Subcommittee of the regional water supply planning process. Seattle, WA. Available online at http://www.climate.tag.washington.edu/ techmemos/Final_Report__12-13-07_.pdf.

[87] Parry, M.L., O Canziani, J.P., Palutikof, C. et al, eds. (2007). Climate change 2007: impacts, adaptation and vulnerability. Contribution of Working Group II to the fourth assessment report of the intergovernmental panel on climate change. New York: Cambridge University Press. Available online at http://www.ipcc.ch/ipccreports/ar4-wg2.htm.

[88] PlaNYC. (2007a). Climate change. Reduce global warming emissions by more than 30%. Available online at http://www.nyc.gov/html/planyc 2030/downloads/ pdf/report_climate_change.pdf.

[89] PlaNYC. (2007b). Water network. Available online at http://www. nyc.gov/html/planyc2030/html/plan/water network.shtml.

[90] PlaNYC. (2008). Sustainable stormwater management plan 2008. Available online at http://www.nyc.gov/html/planyc2030/downloads/ pdf/sustainable_stormwaterplan.pdf.

[91] PlaNYC. (2008–2010). Progress reports for 2008 through 2010. Available online at http://www.nyc.gov/html/planyc2030/html/ downloads/ download.shtml.

[92] Polebitski, A., Traynham, L. & Palmer, R.N. (2007a). Technical memorandum #4: approach for developing climate impacted meteorological data and its quality assurance/quality control. A report prepared by the Climate Change Technical Subcommittee of the Regional Water Supply Planning Process. Seattle, WA. Available online at http://www.govlink.org/regional-water-planning/tech-committees/ climate-change/UWreports/TechnicalMemo4(12- 1 3-07).pdf.

[93] Polebitski, A. Wiley, M.W. & Palmer, R.N. (2007b). Technical memorandum #2: methodology for downscaling meteorological data for evaluating climate change. A report prepared by the Climate Change Technical Subcommittee of the Regional Water Supply Planning Process. Seattle, WA. Available online at http://www.govlink.org/ regional-water-planning/tech-committees/ climate-change/UWreports/ TechnicalMemo2(12- 1 3-07).pdf.

[94] Rosenzweig, C. Major, D.C. Demong, K. et al. (2007). Managing climate change risks in New York City's water system: assessment and

adaptation planning. *Mitigat Adapt Strat Global Change, 12*(8), 1391–1409.

[95] RWSP (Regional Water Supply Planning). (2010). Regional watter supply planning. Seattle, WA. Available online at_http://www.gov link.org/regional-water-planning/index.htm.

[96] Smit, B. Pilifosova, O. Burton, I. et al. (2001). Adaptation to climate change in the context of sustainable development and equity. In: McCarthy, J; Canziana, N; Leary, NA; et al, eds. Climate change 2001: impacts, adaptation, and vulnerability, chapter 18. New York: Cambridge University Press; pp 877–9 12. Available online at http://www.grida.no/climate/ipcc tar/wg2/pdf/wg2TARchap1 8.pdf.

[97] Spartanburg Water. (2009). Lake Blalock buffer management plan. Public works of the city of Spartanburg, SC. September 2009. Accessed March 10. Available online at http://www.sws-sssd.org/pdfs/lbbmp.pdf.

[98] Spartanburg Water. (2010a). Spartanburg water system/Spartanburg sanitary sewer district service areas. Public works of the city of Spartanburg, SC. Accessed March 10. Available online at http://www.sws-sssd.org/watersewer/index.html.

[99] Spartanburg Water. (2010b). Spartanburg lakes. Public works of the city of Spartanburg, SC. [Accessed March 10] Available online at http://www.sws-sssd.org/lakes/index.html.

[100] Spartanburg Water. (2010c). Drought Committee summary. Received on April 22 via e-mail from Rebecca West, Deputy General Manager of Technical and Engineering Services.

[101] Spartanburg Water. (2010d) SWS 10 year Capital Improvement Plan, August 2009. Received on April 22 via email from Rebecca West, Deputy General Manager of Technical and Engineering Services.

[102] Spartanburg Water. (2010e). Operations Capital Improvement Plan items table. Received on April 22 via e-mail from Rebecca West, Deputy General Manager of Technical and Engineering Services.

[103] SPU (Seattle Public Utilities). (2006a). SPU 2007 water system plan. City of Seattle, Washington. November. Available online at http://www.cityofseattle.net/util/About SPU/Water System/Plans/ 2007WaterSystemPlan/index.asp.

[104] SPU (Seattle Public Utilities). (2006b). Water shortage contingency plan: supplement to the Seattle public utilities 2007 water system plan. City of Seattle, Washington. July. Available online at http://www.cityof

seattle.net/util/stellent/groups/public/@spu/@rmb/@resplan/@plancip/documents/webcontent/wat ershor 200312021018123. pdf.

[105] SPU (Seattle Public Utilities). (2008). South Fork Tolt watershed management plan. Prepared by Tetra Tech Infrastructure Group and Triangle Associates. City of Seattle, WA. November. Available online at http://www.cityofseattle.net/util/stellent/groups/public/@spu/@ssw/documents/webcontent/spu01 004082.pdf.

[106] SPU (Seattle Public Utilities). (2010a). Water system history. City of Seattle, WA. Available online at http://www.seattle.gov/util/About_SPU/Water_System/History_&_Overview/WATERSYST_200312020908156.as_p.

[107] SPU (Seattle Public Utilities). (2010b). History of the watershed. City of Seattle, WA. Available online at http://www.seattle.gov/util/About_SPU/Water_System/Water_Sources_&_Treatment/Cedar_River_Watershed/COS 001714.asp.

[108] SPU (Seattle Public Utilities). (2010c). Tolt River watershed. City of Seattle, WA. Available online at http://www.seattle.gov/util/About_SPU/ Water_System/Water_ Sources_&_Treatment/Tolt_River_Watershed/index._asp.

[109] SPU (Seattle Public Utilities). (2010d). Tolt treatment facility. City of Seattle, WA. Available online at http://www.seattle.gov/util/About SPU/Water System/Water Quality/Tolt Treatment Facility/index.asp.

[110] SPU (Seattle Public Utilities). (2010e). Demographics and water use statistics. City of Seattle, WA. Available online at http://www.seattle.gov/util/About_SPU/Water_System/History_&_ Overview/DEMOGRAPHI_200312020908145.a_sp.

[111] SRCC (Southeast Regional Climate Center). (2010). Landrum, South Carolina 1971–2000 Monthly Climate Summary. Accessed May 3. Available online at http://www.sercc.com/.

[112] Stewart, I.T., Cayan, D.R. & Dettinger, M.D. (2004). Changes in snowmelt runoff timing in western North America under a 'business as usual' climate change scenario. *Clim Change*, *62*, 217–232. Available online at http://meteora.ucsd.edu/cap/stewart clch.pdf.

[113] Stratus Consulting and MWH Global. (2009). Implications of Climate Change for Adaptation by Wastewater and Stormwater Agencies. Water Environment Research Foundation Report No. CC2R08.

[114] Towey, A. (2010). Personal communication from A. Towey, EBMUD to Viktoria Zoltay, Abt Associates on April 9, 2010.

[115] U.S. EPA (Environmental Protection Agency). (2002). Cases in water conservation: how efficiency programs help water utilities save money and avoid costs. Available online at http://www.epa.gov/WaterSense/docs/utilityconservation_508.pdf.

[116] U.S. EPA (Environmental Protection Agency). (2009 a). Proceedings of the first national expert and stakeholder workshop on water infrastructure sustainability and adaptation to climate change. Office of Water, Washington, DC; EPA 600/R-09/01 0. Available online at http://www.epa.gov/nrmrl/wswrd/wqm/wrap/workshop.html.

[117] U.S. EPA (Environmental Protection Agency). (2009b). New York City watershed. Region 2. Available online at http://www.epa.gov/Region2/water/nycshed/.

[118] U.S. EPA (U.S. Environmental Protection Agency). (2010). Climate change vulnerability assessments: A review of water utility practices. Office of Water, Washington, DC; EPA/800/R-10/001. Available at http://water.epa.gov/scitech/climatechange/upload/Climate-Change-Vulnerability-Assessments-Sept-2010.pdf.

[119] USGCRP (U.S. Global Change Research Program). (2009). Global climate change impacts in the United States. Available online at http://www.globalchange.gov/publications/reports/scientific-assessments/us-impacts.

[120] USGS (U.S. Geological Survey). (2009). Taste and odor occurance in Lake William C. Bowen and Municipal Reservoir #1, Spartanburg County, South Carolina. USGS Fact Sheet 2009-3 043. Available online at http://pubs.usgs.gov/fs/2009/3043/.

[121] UW-CIG (University of Washington Climate Impacts Group). (2010a). About Pacific Northwest climate. Joint Institute for the Study of the Atmosphere and Ocean, University of Washington, Seattle. Available online at http://cses.washington.edu/cig/pnwc/pnwc.shtml.

[122] UW-CIG (University of Washington Climate Impacts Group). (2010b). Climate change. Joint Institute for the Study of the Atmosphere and Ocean, University of Washington, Seattle. Available online at http://cses.washington.edu/cig/pnwc/cc.shtml.

[123] UW-CIG (University of Washington Climate Impacts Group). (2010c). Case study: Seattle public utilities. Joint Institute for the Study of the Atmosphere and Ocean, University of Washington, Seattle. Available online at http://cses.washington.edu/cig/fpt/casestudyspu.shtml.

[124] WA DOE (Washington State Department of Ecology). (2001). Appendix B: Summary of major ASR projects in Washington. Appendix to ASR

2001 Report to the Legislature. http://www.ecy.wa.gov/programs/wr/asr/images/pdf/ASRLegRptApdx.pdf.

[125] Wallis, M.J., Ambrose, M.R., & Chan, C.C. (2008). Climate change: charting a water course in an uncertain future. *JA WWA, 100*(6), 70–79. http://www.ebmud.com/sites/default/files/pdfs/Journal-06-08_0.pdf.

[126] Water-technology.. (2010). Croton water filtration plant, New York, USA. Available online at http://www.watertechnology.net/projects/crotonfiltration/.

[127] West, R. (2010). Deputy General Manager of Engineering & Technical Services, *Spartanburg Water*. Personal communication on April 1. 2010

[128] WRCC (Western Regional Climate Center). (2010). Western U.S. climate historical summaries. Available online at http://www.wrcc.dri.edu/Climsum.html.

[129] WUCA (Water UtilityClimate Alliance). (2010). About us. Accessed May 3, 2010. Available online at http://www.wucaonline.org/html/about us.html.

End Notes

[1] Existing supplies were estimated to be sufficient during normal and wet years.

[2] For more on the difference between "top-down" and "bottom-up" approaches, see Miller and Yates (2006), Freas et al. (2008), and Stratus Consulting and MWH Global (2009).

[3] For information on climate change and its projected impacts on water resources, see IPCC, 2008 and USGCRP, 2009.

[4] Existing supplies were estimated to be sufficient during normal and wet years.

[5] The existing system was composed of the existing components of the water supply system and projects that were expected to be online by 2010 (i.e., Bayside groundwater and Freeport surface water, see Section 2.2.1).

[6] The 4°C change is based on projected increases from 1980 to 2040 or 2.15°C from 2005 to 2040.

[7] Alum serves as a coagulant, precipitating suspended solids from raw water, reducing objectionable color and turbidity.

[8] High-level storm sewers alleviate pressure on the combined sewer system by capturing about 50% of the rainfall before it enters combined sewer pipes and diverting it directly into waterways. Because such systems require a separate pipe and outlet to a water body, they are generally cost-effective only near the water's edge.

[9] Note that the A2 emission scenario is relatively high by the second half of the twenty-first century, while the B1 scenario has the lowest level of greenhouse gas emissions of the SRES family.

[10] Firm yield is a calculation of how much water can be guaranteed from a water system, in this case, based on a 98% reliability standard.

[11] Tier 1 did include one structural adjustment—the raising of one overflow dike.

[12] Changes include, among others (1) allowing Chester Morse Lake to refill to 1,563 ft (versus 1,560 ft), increasing Cedar River watershed storage by 12%; and (2) allowing South Fork

Tolt Reservoir drawdown to 1,690 ft (versus 1,710 ft), increasing Tolt watershed storage by 18%.

[13] The 7Q10 is the lowest streamflow for 7 consecutive days that is expected to occur once every 10 years.

[14] Hydroelectric power is used to pump water into the distribution or storage system.

[15] For more on the difference between "top-down" and "bottom-up" approaches, see Miller and Yates (2006), Freas et al. (2008), and Stratus Consulting and MWH Global (2009).

INDEX

S

T

U

V

W

Y